PRACTICE WORKB

Algebra 2

HOLT, RINEHART AND WINSTON

A Harcourt Classroom Education Company

Austin • New York • Orlando • Atlanta • San Francisco • Boston • Dallas • Toronto • London

To the Student

Algebra 2 Practice Workbook is designed to provide additional practice of the skills taught in each lesson of your textbook. On each page you will practice the skills from one particular lesson. There are approximately 10 to 50 practice items on each page. These items include practice of both the basic skills and the mathematical applications taught in the lesson.

Photo Credit
Front Cover: Tom Paiva/FPG International.

Printed in the United States of America

ISBN 0-03-054084-4

20 21 018 09 08 07

Table of Contents

Chapter 1   Data and Linear Representations   1

Chapter 2   Numbers and Functions   9

Chapter 3   Systems of Linear Equations & Inequalities   16

Chapter 4   Matrices   22

Chapter 5   Quadratic Functions   27

Chapter 6   Exponential and Logarithmic Functions   35

Chapter 7   Polynomial Functions   42

Chapter 8   Rational Functions & Radical Functions   47

Chapter 9   Conic Sections   55

Chapter 10   Discrete Mathematics: Counting Principles
             and Probability   61

Chapter 11   Discrete Mathematics: Series and Patterns   68

Chapter 12   Discrete Mathematics: Statistics   76

Chapter 13   Trigonometric Functions   82

Chapter 14   Further Topics in Trigonometry   88

Practice

1.1 Tables and Graphs of Linear Equations

State whether each equation is linear.

1. $y = 2x + 1$ _____

2. $y = \frac{3}{4}x$ _____

3. $y = 3.5x - 7x^2$ _____

4. $y = -7 + \frac{2}{7}x$ _____

5. $y = 6.7 - 6.7x^2$ _____

6. $y = 3x + 7x$ _____

7. $y = 9x - \frac{3}{4}$ _____

8. $y = \frac{2}{3}x + \frac{2}{5}x^2$ _____

9. $y = \frac{3}{8}x - 6$ _____

10. $y = 2x + x^2$ _____

11. $y = -\frac{4}{5}x^2$ _____

12. $y = 7x + 2$ _____

Graph each linear equation.

13. $y = \frac{2}{3}x + 4$

14. $y = \frac{4}{5}x$

15. $y = 3x + 2$

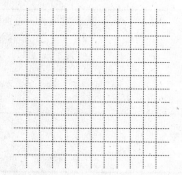
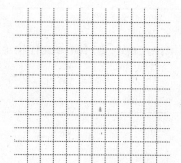
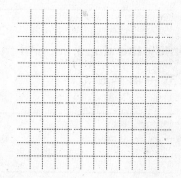

Determine whether each table represents a linear relationship between x and y. If so, write the next ordered pair that would appear in the table.

16.

x	y
2	11
3	13
4	15
5	17

17.

x	y
0	4
1	8
2	12
3	16

18.

x	y
−1	5
−2	11
−3	21
−4	35

19.

x	y
0	−3
−1	−8
−2	−13
−3	−18

20.

x	y
0	3
2	−5
4	−13
6	−21

21.

x	y
1	−6
2	−8
3	−10
4	−12

Practice

1.2 Slopes and Intercepts

Write the equation in slope-intercept form for the line that has the indicated slope, *m*, and *y*-intercept, *b*.

1. $m = 2, b = -5$ _____

2. $m = 3, b = 1$ _____

3. $m = -4, b = 3$ _____

4. $m = \frac{4}{5}, b = -\frac{2}{5}$ _____

5. $m = \frac{1}{6}, b = 3$ _____

6. $m = \frac{1}{4}, b = 4$ _____

Find the slope of the line containing the indicated points.

7. $(3, 0)$ and $(-3, 4)$

_____

8. $\left(-1, -\frac{1}{5}\right)$ and $\left(\frac{2}{3}, \frac{3}{4}\right)$

_____

9. $(2, 6)$ and $(1, 5)$

_____

10. $(-1, -5)$ and $(2, 4)$

_____

Identify the slope, *m*, and the *y*-intercept, *b*, for each line.

11. $3x + 4y = 6$

_____

12. $\frac{3}{4}x + 2y = -3$

_____

13. $-2x - y = 4$

_____

14. $15x + 5y = -35$

_____

Write an equation in slope-intercept form for each line.

15.

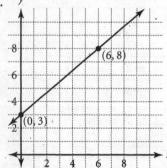

16.

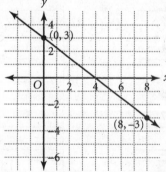

_____

_____

Practice

1.3 *Linear Equations in Two Variables*

Write an equation for the line containing the indicated points.

1. (2, 4) and (3, 5) _____
2. (−1, 3) and (3, −1) _____
3. (3, 1) and $\left(\frac{1}{2}, \frac{3}{2}\right)$ _____
4. (2, 0) and (−6, 4) _____
5. (−1, −4) and (−2, 5) _____
6. $\left(\frac{1}{2}, \frac{3}{2}\right)$ and $\left(-2, -\frac{1}{2}\right)$ _____

Write an equation in slope-intercept form for the line that has the indicated slope, *m*, and contains the given point.

7. $m = 1$ and (3, 3) _____
8. $m = -\frac{1}{2}$ and (4, 6) _____
9. $m = \frac{3}{4}$ and (4, −2) _____
10. $m = 4$ and (4, 3) _____
11. $m = -2$ and (−2, 3) _____
12. $m = -\frac{1}{4}$ and (8, 6) _____

Write an equation in slope-intercept form for the line that contains the given point and is parallel to the given line.

13. (1, 4); $y = -3x + 2$ _____
14. (−2, 3); $y = -4x + 2$ _____
15. (4, −2); $y = \frac{3}{4}x + \frac{1}{4}$ _____
16. (−6, 3); $y = 2x + 2$ _____
17. (2, −1); $y = -3x - 6$ _____
18. (3, −4); $y = 4x - 3$ _____
19. (2, −2); $y = -\frac{1}{2}x - 3$ _____
20. (1, −1); $y = 3x - 2$ _____
21. (2, −2); $y = \frac{1}{2}x + 3$ _____
22. (1, 0); $y = -3x - 2$ _____

Write an equation in slope-intercept form for the line that contains the given point and is perpendicular to the given line.

23. (2, 4); $y = \frac{1}{2}x + 3$ _____
24. (6, −4); $y = 3x - \frac{3}{4}$ _____
25. (6, −7); $y = -2x - 5$ _____
26. (2, −5); $y = 2x - 4$ _____
27. $\left(3, \frac{11}{4}\right)$; $y = 4x + 6$ _____
28. (3, 5); $y = -x - 1$ _____
29. $\left(1, \frac{2}{3}\right)$; $y = \frac{3}{4}x + 3$ _____
30. (1, 4); $y = -\frac{3}{4}x - 4$ _____
31. (3, −1); $y = 3x + \frac{3}{4}$ _____
32. $\left(-1, -\frac{7}{2}\right)$; $y = 4x - 3$ _____

Practice

1.4 Direct Variation and Proportion

In Exercises 1–8, *y* varies directly as *x*. Find the constant of variation, and write an equation of direct variation that relates the two variables.

1. $y = -10$, for $x = 2$ _____

2. $y = 7$, for $x = 3$ _____

3. $y = 4$, for $x = -3$ _____

4. $y = 3.2$, for $x = 12.8$ _____

5. $y = -2$, for $x = -7$ _____

6. $y = 5$, for $x = 6$ _____

7. $y = \frac{2}{3}$, for $x = \frac{1}{3}$ _____

8. $y = -\frac{3}{5}$, for $x = \frac{1}{5}$ _____

Solve each proportion for the variable. Check your answers.

9. $\frac{x}{4} = \frac{9}{12}$ _____

10. $\frac{x+2}{3} = \frac{3x}{6}$ _____

11. $\frac{y}{6} = \frac{18}{24}$ _____

12. $\frac{2x-5}{10} = \frac{3x}{20}$ _____

13. $\frac{z}{10} = \frac{60}{240}$ _____

14. $\frac{3y}{10} = \frac{y-1}{6}$ _____

15. $\frac{x+2}{5} = \frac{5}{25}$ _____

16. $\frac{5z}{7} = \frac{z+3}{14}$ _____

Determine whether the values in each table represent a direct variation. If so, write an equation for the variation. If not, explain why not.

17.

x	y
−2	−6
−1	−3
0	0
1	3
2	6

18.

x	y
5	49
4	28
3	20
2	5
1	2

19.

x	y
1	2
3	6
5	10
7	14
9	18

_____

_____

Practice

1.5 Scatter Plots and Least-Squares Lines

Create a scatter plot of the data in the table below. Describe the correlation. Then find an equation for the least-squares line.

1.

x	y
0	4
2	10
6	22
8	28

2.

x	y
0	1.9
2	10
6	-20.15
8	-27.45

3.

x	y
1	-2
2	-18
7	-26
9	-34

_____

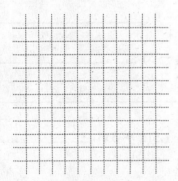

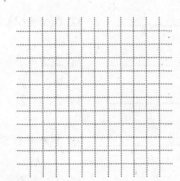

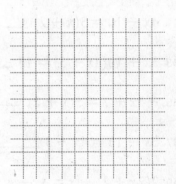

A baseball player has played baseball for several years. The following table shows his batting average for each year over a 10-year period.

1988	1989	1990	1991	1992	1993	1994	1995	1996	1997
0.250	0.258	0.262	0.280	0.272	0.278	0.285	0.292	0.316	0.320

In Exercises 4–6, refer to the table above.

4. Enter the data in a graphics calculator, and find the equation of the least-squares line. _____

5. Find the correlation coefficient, r, to the nearest tenth. _____

6. Use the least-squares line to predict the baseball player's batting average in 1999. _____

Practice

1.6 Introduction to Solving Equations

Solve each equation.

1. $4x + 4(2x - 1) = 20$ _____

2. $4x + 20 = 5(x + 3)$ _____

3. $5x + 15 = 10(x - 3)$ _____

4. $2x + 5 = 17$ _____

5. $3x - 4 = 4(3x - 19)$ _____

6. $3(2x - 4) = 3x - 5(x + 1)$ _____

7. $-0.4x - 6(3x - 2) = 48.8$ _____

8. $2(x + 3) = 5x + 15$ _____

9. $2x + 1 = 7 - 10x$ _____

10. $5x - 3 = 15 - 4x$ _____

11. $4x - 10 = 3(x + 2)$ _____

12. $6(x + 2) = 5x - 9$ _____

13. $5x + 10(4x + 3) = 15$ _____

14. $2(x + 3) = 5(x - 3)$ _____

15. $-4x + 7 = 5(x + 2)$ _____

16. $7x = 2(x - 3)$ _____

17. $5x - 15 = 4x + 3$ _____

18. $5(x + 0.5) = -1.5(x + 3x)$ _____

19. $2(2x + 2) + x = 3x - 4$ _____

20. $2x = 3(x + 2)$ _____

21. $2x + 4(3x + 6) = 12$ _____

22. $2x + 2(2x - 3) = -3$ _____

Solve each literal equation for the indicated variable.

23. $L \times W \times D = V$, for W _____

24. $C = 2\pi r$, for r _____

25. $V_1 P_1 = V_2 P_2$, for P_1 _____

26. $q = q_p \times D \times Q$, for q_p _____

27. $T = T_o - a(z - z_0)$, for a _____

28. $A = (a + b)h$, for h _____

Practice

1.7 *Introduction to Solving Inequalities*

Write an inequality that describes each graph.

1. [number line from −6 to 6, closed point at −3, shaded right] → x _____

2. [number line from −6 to 6, open point at 2, shaded right] → x _____

3. [number line from −6 to 6, closed point at 2, shaded left] → x _____

Solve each inequality, and graph the solution on a number line.

4. $\dfrac{2x-1}{3} \geq x + 1$ _____

   [number line] → x

5. $7x - 15 \leq -2(x + 3)$ _____

   [number line] → x

6. $3x + 4 < 3(x + 2)$ _____

   [number line] → x

7. $-5(x + 2) \leq 3x + 6$ _____

   [number line] → x

Graph the solution of each compound inequality on a number line.

8. $5x - 2 < 3 \text{ or } 2x - 6 < 4$ _____

   [number line] → x

9. $150 < \dfrac{t + 738}{6} \text{ and } \dfrac{t + 738}{6} < 155$ _____

   [number line marked 150, 160, 170, 180, 190, 200] → x

Practice

1.8 Solving Absolute-Value Equations and Inequalities

Solve each equation. Graph the solution on a number line.

1. $|x + 3| = 5$ _____

2. $|x - 4| = 6$ _____

3. $|2x + 5| = 7$ _____

4. $|5x + 3| = 12$ _____

5. $|3x + 12| = 18$ _____

Solve each inequality. Graph the solution on a number line.

6. $|5x + 2| < 7$ _____

7. $|6x - 4| < 3$ _____

8. $|5x - 6| < 5$ _____

9. $|3x + 6| > 15$ _____

10. $|4x - 5| \geq 15$ _____

Practice
2.1 Operations With Numbers

Classify each number in as many ways as possible.

1. $\frac{13}{17}$ _____

2. $\sqrt{91}$ _____

3. $3.12112111211112\ldots$ _____

4. 801.35 _____

5. $-\sqrt{900}$ _____

6. $501.\overline{07}$ _____

State the property that is illustrated in each statement.
Assume that all variables represent real numbers.

7. $75 + (-75) = 0$ _____

8. $181 \cdot 1 = 181$ _____

9. $-2 + (33 + 18) = (-2 + 33) + 18$ _____

10. $\frac{54}{k} \cdot \frac{k}{54} = 1$, where $k \neq 0$ _____

11. $47y \cdot 3x = 3x \cdot 47y$ _____

12. $14(x + 91) = 14x + 14(91)$ _____

13. $\frac{7}{8} + 0 = \frac{7}{8}$ _____

Evaluate each expression by using the order of operations.

14. $-2 \cdot 4^2 - 1$ _____

15. $52 \div (2 + 11)$ _____

16. $27 + 8 \cdot 2$ _____

17. $45 - 16 \div 8$ _____

18. $13 \times 3 + 2 \times 5$ _____

19. $12 + 8^2 \div 4$ _____

20. $\frac{150 - 38}{4} - 4 + 2$ _____

21. $(13 - 7)^2 \div 5$ _____

22. $(77 - 50) - (13 - 42)$ _____

23. $7 \cdot 12 + 30 \div 5$ _____

Practice
2.2 Properties of Exponents

Evaluate each expression.

1. 32^0 _____

2. $-(15^{-1})$ _____

3. $(2 \cdot 3)^2$ _____

4. $(-3^4 3^5)^0$ _____

5. $(-217)^1$ _____

6. $\left(\frac{3}{5}\right)^2$ _____

7. $\left(\frac{1}{5}\right)^{-2}$ _____

8. $\left(\frac{1}{2}\right)^{-5}$ _____

9. $32^{\frac{1}{5}}$ _____

10. $\left(\frac{1}{2}\right)^{-4}$ _____

11. $64^{\frac{5}{6}}$ _____

12. $(-27)^{\frac{2}{3}}$ _____

Simplify each expression, assuming that no variable equals zero.
Write your answer with positive exponents.

13. $d^3 d^{-4}$ _____

14. $w^3 y^4 z \cdot w y^{-2} z$ _____

15. $k^{-11} k^3$ _____

16. $(x^7)^2$ _____

17. $\dfrac{z^{15}}{z^{-2}}$ _____

18. $\left(\dfrac{1}{x^{-7}}\right)^{-5}$ _____

19. $\dfrac{y^{14} z^5}{y^9 z^4}$ _____

20. $\dfrac{w^{21} w^{-12}}{w^9}$ _____

21. $(3x^3 y^5)^4$ _____

22. $\left(-2a^3 bc^6\right)^4$ _____

23. $(5a^2 b^3)^3$ _____

24. $\left(\dfrac{wz^4}{x^2}\right)^{-2}$ _____

25. $\left(\dfrac{a^{-2}}{b^{-3}}\right)^{-2}$ _____

26. $\left(\dfrac{w^6}{k}\right)^3$ _____

27. $\left(\dfrac{m^{-2} p^2}{2mp^3}\right)^{-4}$ _____

28. $\left(\dfrac{xy^3 z^2}{z^{-2}}\right)^{-1}$ _____

29. $\left[\dfrac{(x^2 y^2)^3}{x^5}\right]^{-1}$ _____

30. $\left(\dfrac{3x}{y}\right)^4 \left[\dfrac{x^{-8}}{(xy)^3}\right]^{-2}$ _____

Practice
2.3 Introduction to Functions

In Exercises 1–8, state whether each relation represents a function.

1.

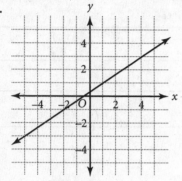

_____

2.

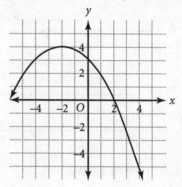

_____

3.

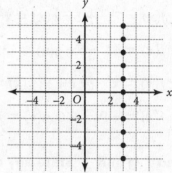

_____

4.

x	-2	-1	0	1	2
y	8	4	0	4	8

_____

5.

x	1	2	3	4	5
y	2	3	2	3	2

_____

6.

x	2	3	2	3	2
y	1	2	3	4	5

_____

7. $\{(1, 5), (0.5, 8), (0, 3)\}$ _____

8. $\{(32, 4), (16, 7), (16, 4)\}$ _____

State the domain and range of each function.

9. $\left\{(-1, -3), (0, 1), \left(\frac{1}{2}, 3\right), \left(\frac{3}{2}, 7\right)\right\}$

_____

10.

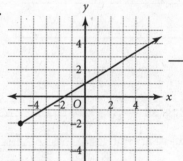

_____

11. $\{(-4.5, \ 6), (3, -1.5), (6.5, -5), (12, -10.5)\}$

_____

12. $\{(-2, 12), (0, 8), (1, 9)(5, 33)\}$

_____

Evaluate each function for the given values of x.

13. $f(x) = 20x - 4$, for $x = -2$ and $x = 8$ _____

14. $f(x) = 5x^2$, for $x = -3$ and $x = 5$ _____

15. $f(x) = 12 - 3x$, for $x = 7$ and $x = -5$ _____

16. $f(x) = 3x^2 - 2$, for $x = 11$ and $x = -4$ _____

17. $f(x) = 3x - x^2$, for $x = 0.5$ and $x = 0$ _____

Practice
2.4 Operations With Functions

Find $f + g$ and $f - g$.

1. $f(x) = 7x^2 + 5x; g(x) = x^2 - 13$ _____

2. $f(x) = 41 - 5x; g(x) = 13x^2$ _____

3. $f(x) = x^2 + \frac{1}{3}x + 9; g(x) = -7x - 7$ _____

4. $f(x) = -9x^2 + 6; g(x) = 12x^2$ _____

Find $f \cdot g$ and $\frac{f}{g}$. State any domain restrictions.

5. $f(x) = 35x + 5; g(x) = 5$ _____

6. $f(x) = x^2 + 25; g(x) = 3x + 17$ _____

7. $f(x) = x^2 + 16; g(x) = x^2 - 16$ _____

Let $f(x) = -2x - 2$ and $g(x) = x + 10$. Find each new function, and state any domain restrictions.

8. $f + g$ _____   9. $f - g$ _____

10. $g - f$ _____   11. $f \cdot g$ _____

12. $\frac{f}{g}$ _____   13. $\frac{g}{f}$ _____

Find $f \circ g$ and $g \circ f$.

14. $f(x) = 3x - 2; g(x) = \frac{1}{3}(x + 2)$ _____

15. $f(x) = 4x; g(x) = x^2 - 1$ _____

16. $f(x) = -x^2 + 1; g(x) = x$ _____

Let $f(x) = 11x$, $g(x) = x^2 - 5$, and $h(x) = 2(x-4)$. Evaluate each composite function.

17. $(f \circ g)(-1)$ _____   18. $(h \circ f)(-2)$ _____   19. $(h \circ g)(2)$ _____

20. $(g \circ h)(4)$ _____   21. $(g \circ f)(0)$ _____   22. $(f \circ h)(5)$ _____

23. $(f \circ g)(0)$ _____   24. $(h \circ h)(-1)$ _____   25. $(f \circ f)(2)$ _____

Practice
2.5 Inverses of Functions

Find the inverse of each relation. State whether the relation is a function and whether its inverse is a function.

1. $\{(-1, -16), (0, -6), (2, 14)\}$ _____

2. $\{(7, 2), (6, 3), (7, 4), (6, 5)\}$ _____

3. $\{(-2, 16), (-1, 1), (1, 1), (2, 16)\}$ _____

4. $\{(-5, 7), (-3, 7), (-1, 7), (1, 7)\}$ _____

5. $\{(-5, 4), (-3, 9), (1, 12), (7, 13)\}$ _____

For each function, find the equation of its inverse. Then use composition to verify that the equation you wrote is the inverse.

6. $f(x) = \frac{1}{3}(x + 1)$

7. $h(x) = \frac{2x+1}{3}$

_____   _____

8. $g(x) = 11x - 4$

9. $f(x) = \frac{1}{2}(x - 2.5)$

_____   _____

10. $g(x) = 8(x + 2)$

11. $h(x) = \frac{x}{4} - 8$

_____   _____

Graph each function, and use the horizontal-line test to determine whether the inverse is a function.

12. $f(x) = x^2 + x$

13. $g(x) = x^3 + 1$

14. $h(x) = x^3 + x$

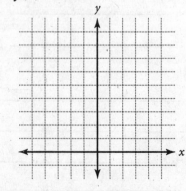   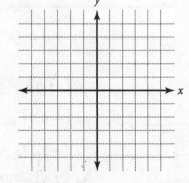   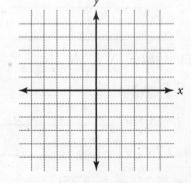

_____   _____   _____

Practice
2.6 Special Functions

Graph each function.

1. $g(x) = \begin{cases} -1 - x & \text{if} < 0 \\ x + 1 & \text{if } x \geq 0 \end{cases}$

2. $f(x) = \begin{cases} -x - 4 & \text{if } x \leq -2 \\ -x + 1 & \text{if } x > -2 \end{cases}$

3. $h(x) = -2[x]$

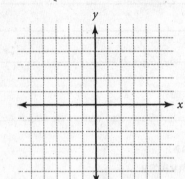
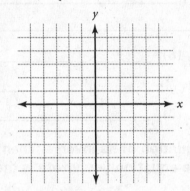
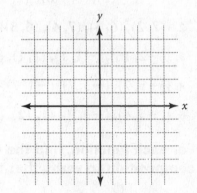

Write the piecewise function represented by each graph.

4.
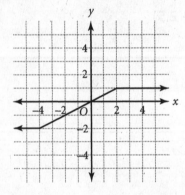

_____

_____

_____

5.
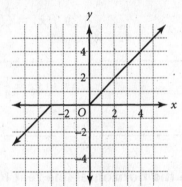

_____

_____

_____

6.
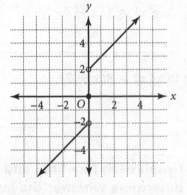

_____

_____

_____

Evaluate.

7. $[-9.23]$ _____

8. $\lceil 31.7 \rceil$ _____

9. $|13.13|$ _____

10. $\lceil -0.9 \rceil$ _____

11. $\lceil 7.8 \rceil + \lceil -1.88 \rceil$ _____

12. $[-2.22] - |-4.5|$ _____

13. $|5.25| - |-3.75|$ _____

14. $\lceil 2.5 \rceil - [2.5]$ _____

15. $-\lceil 12.95 \rceil - [6.3]$ _____

16. $-|-3| - [4.9]$ _____

Practice
2.7 A Preview of Transformations

Identify each transformation from the parent function $f(x) = x^2$ to g.

1. $g(x) = (x + 7.5)^2$ _____

2. $g(x) = x^2 + 7.5$ _____

3. $g(x) + (52x)^2$ _____

4. $g(x) = -2x^2$ _____

5. $g(x) = 14x^2 + 6$ _____

6. $g(x) = 12(x - 7)^2$ _____

Identify each transformation from the parent function $f(x) = \sqrt{x}$ to g.

7. $g(x) = \sqrt{x + 21}$ _____

8. $g(x) = 17\sqrt{x}$ _____

9. $g(x) = \sqrt{\frac{1}{2}x}$ _____

10. $g(x) = \sqrt{x} + 13.7$ _____

11. $g(x) = -3\sqrt{x}$ _____

12. $g(x) = 41\sqrt{x - 8}$ _____

Write the function for each graph described below.

13. the graph of $f(x) = x^3$ reflected across the x-axis _____

14. the graph of $f(x) = x^5$ translated 7 units to the left _____

15. the graph of $f(x) = x^4$ stretched horizontally by a factor of 26 _____

16. the graph of $f(x) = |x|$ compressed vertically by a factor of $\frac{1}{12}$ _____

17. the graph of $f(x) = \frac{2}{3}x + 9$ reflected across the y-axis _____

18. the graph of $f(x) = x^3$ translated 33 units down _____

Practice

3.1 Solving Systems by Graphing or Substitution

Graph and classify each system. Then find the solution from the graph.

1. $\begin{cases} y = x + 4 \\ y + x = 4 \end{cases}$

2. $\begin{cases} \frac{1}{2}x + y = 2 \\ 2y + x = 4 \end{cases}$

3. $\begin{cases} 3x + 4y = -7 \\ 2x + y = -3 \end{cases}$

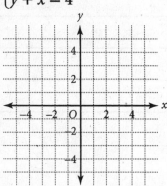

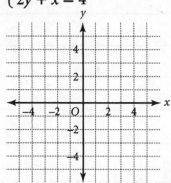

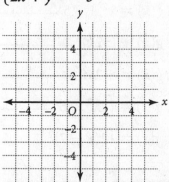

_____ _____ _____

4. $\begin{cases} y = -4x + 10 \\ 2x + \frac{1}{2}y = 6 \end{cases}$

5. $\begin{cases} 5x - y = 2 \\ 2x - y = -1 \end{cases}$

6. $\begin{cases} x + \frac{1}{3}y = 3 \\ 3x - y = -3 \end{cases}$

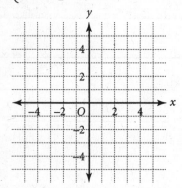

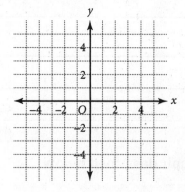

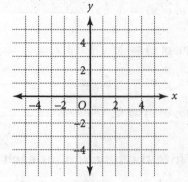

_____ _____ _____

Use substitution to solve each system of equations. Check your solution.

7. $\begin{cases} y + 2x = 11 \\ x + y = 5 \end{cases}$

8. $\begin{cases} x - 10y = 2 \\ x - 6y = 6 \end{cases}$

9. $\begin{cases} 8x = y \\ 2x + y = 5 \end{cases}$

_____ _____ _____

10. $\begin{cases} 3x + y = -4 \\ \frac{1}{2}x + y = 6 \end{cases}$

11. $\begin{cases} x + 2y = 2 \\ 2x + 3y = -1 \end{cases}$

12. $\begin{cases} 3x + 4y = 11 \\ 2x + 4y = 8 \end{cases}$

_____ _____ _____

Practice

3.2 Solving Systems by Elimination

Use elimination to solve each system of equations. Check your solution.

1. $\begin{cases} -2x + 9y = -13 \\ 6x - 3y = 15 \end{cases}$

2. $\begin{cases} \frac{2}{3}x - 3y = \frac{1}{5} \\ 2x - 9y = 4 \end{cases}$

3. $\begin{cases} 7y - x = 8 \\ x - y = 4 \end{cases}$

4. $\begin{cases} 4x + y = 12 \\ 3x + \frac{1}{4}y = 9 \end{cases}$

5. $\begin{cases} 5x + 9y = -7 \\ 2x + 3y = -1 \end{cases}$

6. $\begin{cases} \frac{1}{2}x + y = 22 \\ 2x + 4y = 11 \end{cases}$

7. $\begin{cases} \frac{2}{3}x - y = -2 \\ 3x + 2y = -35 \end{cases}$

8. $\begin{cases} 6x - y = 26 \\ 3x - \frac{1}{2}y = 13 \end{cases}$

9. $\begin{cases} \frac{1}{2}x + \frac{3}{4}y = 10 \\ 2x - y = 8 \end{cases}$

10. $\begin{cases} x - 9y = -13 \\ 2x + y = -7 \end{cases}$

11. $\begin{cases} 13x + 7y = 19 \\ 9x - 2y = 20 \end{cases}$

12. $\begin{cases} 5x + 2y = -9 \\ y - 3x = 12 \end{cases}$

13. $\begin{cases} 11x - 4y = 19 \\ 3x - 2y = 7 \end{cases}$

14. $\begin{cases} 3x - 2y = 31 \\ 3x + 2y = -1 \end{cases}$

15. $\begin{cases} 3x + 5y = 4 \\ 5x + 7y = 6 \end{cases}$

Use any method to solve each system of linear equations. Check your solution.

16. $\begin{cases} y = 5x + 2 \\ y = 2x + 5 \end{cases}$

17. $\begin{cases} y = 6x \\ 2x + 5y = 16 \end{cases}$

18. $\begin{cases} 4x + y = 9 \\ 2y = -8x + 18 \end{cases}$

19. $\begin{cases} x + y = 1 \\ 2x + 5y = 4 \end{cases}$

20. $\begin{cases} y - 7 = x \\ x + y = 13 \end{cases}$

21. $\begin{cases} 2x - 3y = 7 \\ \frac{2}{3}x - y = 9 \end{cases}$

22. $\begin{cases} 12x - y = 2 \\ 4x + 3y = 4 \end{cases}$

23. $\begin{cases} 9x - 5y = -4 \\ 3x + 2y = 6 \end{cases}$

24. $\begin{cases} 5x + 3y = 46 \\ 2x + 5y = 7 \end{cases}$

Practice

3.3 Linear Inequalities in Two Variables

Graph each linear inequality.

1. $y < -x$

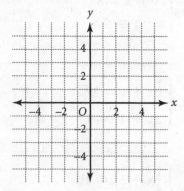

2. $y \geq -3x - 2$

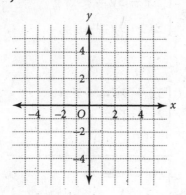

3. $y > \frac{1}{4}x + 1$

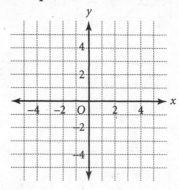

4. $2x + y \leq -3$

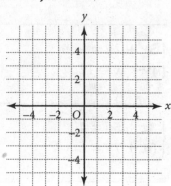

5. $x - y > 4$

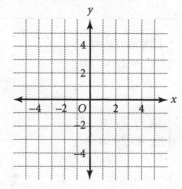

6. $x \geq -1.5$

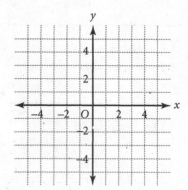

7. Sheila earns a basic wage of $8 per hour. Under certain conditions, she is paid $12 per hour. The most that she can earn in one week is $400.

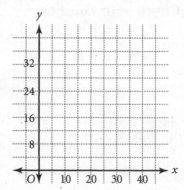

a. Write an inequality that describes her total weekly wages for x hours at $8 per hour and for y hours at $12 per hour.

_____

b. Graph the inequality on the grid at right.

c. What is the maximum number of hours that Sheila can work for $8 per hour? for $12 per hour?

_____

Algebra 2

Practice
3.4 *Systems of Linear Inequalities*

Graph each system of linear inequalities.

1. $\begin{cases} y \le -x \\ y \ge 2x - 4 \end{cases}$

2. $\begin{cases} y > -3x - 3 \\ y \le \frac{1}{3}x - 1 \end{cases}$

3. $\begin{cases} y \le 4 \\ y \ge -3 \\ y > 3x - 2 \end{cases}$

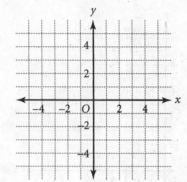 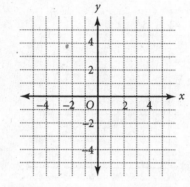 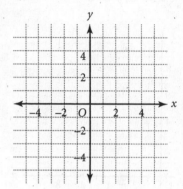

Write the system of inequalities whose solution is graphed.

4.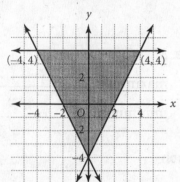

5.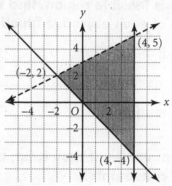

6.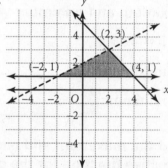

_____   _____   _____

7. During the summer, Ryan works 30 hours or less per week mowing lawns and delivering newspapers. He earns $6 per hour mowing lawns and $7 per hour delivering papers. Ryan would like to earn at least $126 per week. Let x be the number of hours mowing, and let y be the number of hours delivering papers. Write a system of inequalities to represent the possible hours and jobs that Ryan can work, and graph this system at right.

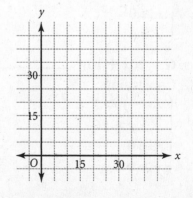

_____

Practice
3.5 Linear Programming

Graph the feasible region for each set of constraints.

1. $\begin{cases} x + y \le 9 \\ 2x - y \le 5 \\ x \ge 0, y \ge 0 \end{cases}$

2. $\begin{cases} 3x + 4y \le 20 \\ y - x \le 3 \\ x \ge 0, y \ge 0 \end{cases}$

3. $\begin{cases} x + 2y \le 16 \\ 3x - 4y \le 12 \\ x \ge 0, y \ge 0 \end{cases}$

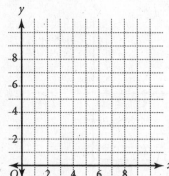
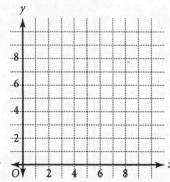
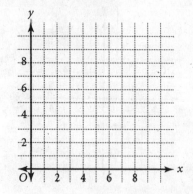

The feasible region for a set of constraints has vertices at (2,0), (10, 1), (8, 5), and (0, 4). Given this feasible region, find the maximum and minimum values of each objective function.

4. $F = 4x + y$

   maximum: _____

   minimum: _____

5. $E = 2x - 3y$

   maximum: _____

   minimum: _____

6. $M = 3y - x$

   maximum: _____

   minimum: _____

Find the maximum and minimum values, if they exist, of each objective function for the given constraints.

7. $P = x + 5y$ _____
   Constraints:
   $\begin{cases} 2x + y \le 10 \\ x - y \le 4 \\ x \ge 0 \\ y \ge 0 \end{cases}$

8. $E = 4x + 8y$ _____
   Constraints:
   $\begin{cases} x + y \le 9 \\ y - x \le 7 \\ x \ge 0 \\ y \ge 0 \end{cases}$

9. $G = 20x + 10y$ _____
   Constraints:
   $\begin{cases} 2x - y \ge 2 \\ x + y \le 10 \\ x \ge 0 \\ y \ge 0 \end{cases}$

10. $F = 12x - 5y$ _____
   Constraints:
   $\begin{cases} 2x - y \le 10 \\ x + 2y \le 10 \\ x \ge 0 \\ y \ge 0 \end{cases}$

Practice
3.6 *Parametric Equations*

Graph each pair of parametric equations for the interval $-3 \le t \le 3$.

1. $\begin{cases} x(t) = \frac{1}{3}t + 4 \\ y(t) = 2t - 1 \end{cases}$

2. $\begin{cases} x(t) = 6 - t \\ y(t) = \frac{1}{2}t + 1 \end{cases}$

3. $\begin{cases} x(t) = \frac{1}{2}t - 5 \\ y(t) = t + 3 \end{cases}$

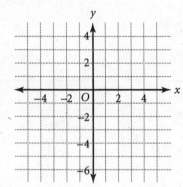
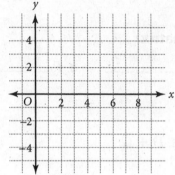
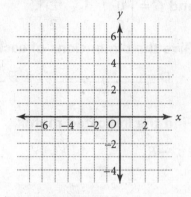

Write each pair of parametric equations as a single equation in *x* and *y*.

4. $\begin{cases} x(t) = 2t + 12 \\ y(t) = t - 8 \end{cases}$

5. $\begin{cases} x(t) = \frac{1}{2}t \\ y(t) = 3t + 2 \end{cases}$

6. $\begin{cases} x(t) = 4 + 2t \\ y(t) = 6 - t \end{cases}$

_____ _____ _____

7. $\begin{cases} x(t) = 10 - 2t \\ y(t) = t + 11 \end{cases}$

8. $\begin{cases} x(t) = 3 - 12t \\ y(t) = 4t - 3 \end{cases}$

9. $\begin{cases} x(t) = 3t \\ y(t) = t^2 + 1 \end{cases}$

_____ _____ _____

An airplane is ascending at a constant rate. Its altitude changes at a rate of 12 feet per second. Its horizontal speed is 150 feet per second.

10. a. Write parametric equations that represent the plane's flight

   path. _____

   b. Graph the equations for the interval $0 \le t \le 30$. Use the grid at right.

11. a. How long will it take the plane to reach an altitude of

   300 ft? _____

   b. How far will the plane travel horizontally in that

   time? _____

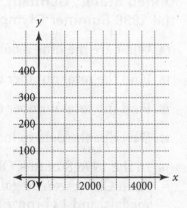

Practice

4.1 Using Matrices to Represent Data

In Exercises 1–13, let $D = \begin{bmatrix} 7 & 2 \\ 3 & 1 \end{bmatrix}$, $E = \begin{bmatrix} 0 & -2 \\ -1 & 4 \end{bmatrix}$, $F = \begin{bmatrix} 2 & 4 & 2 \\ 3 & 3 & 6 \\ 7 & 0 & -5 \end{bmatrix}$,

and $G = \begin{bmatrix} 3 & -3 & -4 \\ 0 & 1 & 2 \\ 6 & -2 & 2 \end{bmatrix}$.

Give the dimensions of each matrix.

1. D _____

2. E _____

3. F _____

4. G _____

Find the indicated matrix.

5. $-F$

6. $D + G$

7. $3E$

_____   _____   _____

8. $D - E$

9. $E - D$

10. $2F + G$

_____   _____   _____

11. $-2G$

12. $F - G$

13. $2D + 3E$

_____   _____   _____

Matrix *M* at right represents the number of medals won by athletes from the United States, Germany, and Russia in the 1996 Summer Olympic Games.

	Gold	Silver	Bronze
United States	44	32	25
Germany	20	18	27
Russia	26	21	16

14. What are the dimensions of matrix M? _____

15. Find the total number of medals won by the United States. _____

16. Find the total number of gold medals won by the three nations. _____

17. Describe the data in location m_{23}. _____

18. In the 1996 Summer Olympic Games, athletes from China won 16 gold medals, 22 silver medals, and 12 bronze medals. Write a new matrix, M', that includes medals for all four countries. _____

Practice

4.2 *Matrix Multiplication*

Find each product, if it exists.

1. $\begin{bmatrix} 2 & 5 \\ -1 & 7 \end{bmatrix}\begin{bmatrix} 2 & 5 \\ -1 & 7 \end{bmatrix}$

2. $[-4 \ \ 0 \ \ 4]\begin{bmatrix} 6 & -1 \\ 0 & 2 \\ 1 & 0 \end{bmatrix}$

3. $\begin{bmatrix} 4 & 2 & 0 \\ 1 & 5 & 2 \end{bmatrix}\begin{bmatrix} 6 & 1 & 6 \\ 9 & 7 & 8 \end{bmatrix}$

4. $\begin{bmatrix} -3 & 4 \\ 2 & 7 \end{bmatrix}\begin{bmatrix} 0 & 2 \\ 1 & 0 \end{bmatrix}$

5. $\begin{bmatrix} 5 \\ 1 \\ 3 \end{bmatrix}[2 \ \ -2 \ \ 1]$

6. $\begin{bmatrix} 4 & 5 \\ 1 & 2 \\ 0 & 3 \end{bmatrix}\begin{bmatrix} 0.5 & 2 \\ 6 & 1 \end{bmatrix}$

7. $\begin{bmatrix} 0 & 2 \\ 1 & 1 \\ -3 & 1 \\ -2 & 4 \end{bmatrix}\begin{bmatrix} 2 \\ 1 \end{bmatrix}$

8. $\begin{bmatrix} 7 & 0 & 1 & 1 \\ -2 & 3 & 5 & 0 \\ 4 & 1 & 1 & -1 \end{bmatrix}\begin{bmatrix} 6 & -1 & 1 \\ -2 & 1 & 4 \end{bmatrix}$

9. $\begin{bmatrix} 0.5 & 4 \\ 6 & -1 \end{bmatrix}\begin{bmatrix} 6 & 0 \\ 2 & 8 \end{bmatrix}$

Matrix *T*, $\begin{bmatrix} -2 & 4 & 2 \\ 0 & 4 & -4 \end{bmatrix}$, represents triangle *XYZ*, which is graphed at right.

10. Find the coordinates of the vertices of the image, triangle *X'Y'Z'*, which is formed by multiplying matrix *T* by the transformation matrix $\begin{bmatrix} \frac{3}{2} & 0 \\ 0 & \frac{3}{2} \end{bmatrix}$.

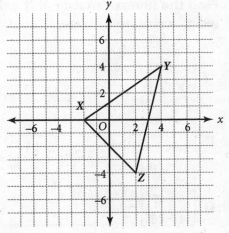

11. Sketch the image, triangle *X'Y'Z'*, on the grid at right above.

12. Describe the transformation.

Practice

4.3 The Inverse of a Matrix

Determine whether each pair of matrices are inverses of each other.

1. $\begin{bmatrix} 4 & -3 \\ -5 & 4 \end{bmatrix}, \begin{bmatrix} 4 & 3 \\ 5 & 4 \end{bmatrix}$ _____

2. $\begin{bmatrix} 5 & -2 \\ -17 & 7 \end{bmatrix}, \begin{bmatrix} 7 & 2 \\ 17 & 5 \end{bmatrix}$ _____

3. $\begin{bmatrix} 6 & 9 \\ 2 & 3 \end{bmatrix}, \begin{bmatrix} 3 & -9 \\ -2 & 6 \end{bmatrix}$ _____

4. $\begin{bmatrix} 3 & -3\frac{2}{3} \\ -4 & 5 \end{bmatrix}, \begin{bmatrix} 15 & 11 \\ 12 & 9 \end{bmatrix}$ _____

5. $\begin{bmatrix} 12 & 5 \\ 14 & 6 \end{bmatrix}, \begin{bmatrix} 3 & -2.5 \\ -7 & 6 \end{bmatrix}$ _____

6. $\begin{bmatrix} \frac{1}{2} & -\frac{1}{8} \\ -\frac{1}{4} & \frac{3}{16} \end{bmatrix}, \begin{bmatrix} 3 & 2 \\ 4 & 8 \end{bmatrix}$ _____

Find the determinant and the inverse of each matrix, if it exists.

7. $\begin{bmatrix} 7 & 5 \\ 4 & 3 \end{bmatrix}$ _____

8. $\begin{bmatrix} 9 & 7 \\ 5 & 4 \end{bmatrix}$ _____

9. $\begin{bmatrix} 8 & 5 \\ 7 & 5 \end{bmatrix}$ _____

10. $\begin{bmatrix} 11 & 6 \\ 7 & 4 \end{bmatrix}$ _____

11. $\begin{bmatrix} 7\frac{1}{2} & 5 \\ 12 & 8 \end{bmatrix}$ _____

12. $\begin{bmatrix} 13 & 3 \\ 16 & 4 \end{bmatrix}$ _____

Find the inverse matrix, if it exists. If the inverse matrix does not exist, write *no inverse*.

13. $\begin{bmatrix} 2 & 1 \\ 3 & 1 \end{bmatrix}$ _____

14. $\begin{bmatrix} 4 & 6 \\ 5 & 7 \end{bmatrix}$ _____

15. $\begin{bmatrix} 6 & 4 \\ -3 & -2 \end{bmatrix}$ _____

16. $\begin{bmatrix} 5 & 3 \\ 2 & 1 \end{bmatrix}$ _____

17. $\begin{bmatrix} \frac{2}{3} & 2 \\ 4 & 12 \end{bmatrix}$ _____

18. $\begin{bmatrix} 1.5 & -2.5 \\ -1 & 2 \end{bmatrix}$ _____

Algebra 2

Practice

4.4 Solving Systems With Matrix Equations

Write the matrix equation that represents each system.

1. $\begin{cases} 3x + y - z = -19 \\ -x - y + 3z = 21 \\ 2x + 2y + z = -7 \end{cases}$

2. $\begin{cases} 5x - 2y + z = 13 \\ -x + 4y - z = -1 \\ 4x - 8y + 3z = 6 \end{cases}$

3. $\begin{cases} 9x - 5y + z = 6 \\ 3x + y - z = 2 \\ 4x - 3y - 2z = -1 \end{cases}$

_____   _____   _____

Write the system of equations represented by each matrix equation.

4. $\begin{bmatrix} 2 & 3 & -1 \\ 3 & 4 & 1 \\ -1 & -1 & 2 \end{bmatrix}\begin{bmatrix} x \\ y \\ z \end{bmatrix} = \begin{bmatrix} 1 \\ 6 \\ 7 \end{bmatrix}$

5. $\begin{bmatrix} 3 & 2 & -1 \\ 2 & 3 & 1 \\ 4 & 4 & 3 \end{bmatrix}\begin{bmatrix} x \\ y \\ z \end{bmatrix} = \begin{bmatrix} -6 \\ 1 \\ 20 \end{bmatrix}$

_____   _____

Write the matrix equation that represents each system, and solve the system, if possible, by using a matrix equation.

6. $\begin{cases} 8x + 7y = 5 \\ 4x - 9y = 65 \end{cases}$

7. $\begin{cases} 7x + 5y = 14 \\ 4x + 3y = 9 \end{cases}$

8. $\begin{cases} 3x - 7y = 25 \\ 5x - 8y = 27 \end{cases}$

_____   _____   _____

9. $\begin{cases} 4x + y + z = 1 \\ 8x - 4y - 7z = 2 \\ 5y + 9z = 3 \end{cases}$

10. $\begin{cases} 3x - 3y + 5z = 13 \\ 5x + 6y - 2z = 10 \\ 7x + 5y = 18 \end{cases}$

11. $\begin{cases} x - 2y - 3z = 3 \\ 3x + y + z = 12 \\ 3x - 2y - 4z = 15 \end{cases}$

_____   _____   _____

12. $\begin{cases} x - 2y + z = 15 \\ 3x + y - 2x = 8 \\ 5x - 10y + 5z = 21 \end{cases}$

13. $\begin{cases} 12x + 7y + z = -5 \\ 3x + 4y + 2z = 3 \\ 5x + 3y - 3z = 12 \end{cases}$

14. $\begin{cases} 8x + y - z = 0 \\ 5x + 2y - 9z = -3 \\ 12x + y + 5z = 8 \end{cases}$

_____   _____   _____

Practice

4.5 *Using Matrix Row Operations*

Write the augmented matrix for each system of equations.

1. $\begin{cases} 4x + 5y + z = 2 \\ 7x + 9y + 2z = 7 \\ x + y + z = 2 \end{cases}$

2. $\begin{cases} 2x + y - 3z = 4 \\ 5x + 6y + z = 6 \\ 7x + 8y - 3z = 2 \end{cases}$

3. $\begin{cases} x + y + z = -1 \\ 2x + 3y + 3z = 4 \\ 6x + 7y + 3z = 8 \end{cases}$

_____   _____   _____

Find the reduced row-echelon form of each matrix.

4. $\begin{bmatrix} 1 & 1 & 0 & \vdots & -1 \\ 0 & 2 & -1 & \vdots & -3 \\ 3 & 0 & -2 & \vdots & 5 \end{bmatrix}$

5. $\begin{bmatrix} 1 & 0 & -2 & \vdots & 0 \\ 0 & 3 & 1 & \vdots & 2 \\ -2 & 1 & 0 & \vdots & 5 \end{bmatrix}$

6. $\begin{bmatrix} 3 & 2 & -1 & \vdots & 7 \\ 1 & -4 & 1 & \vdots & -9 \\ 5 & 0 & -3 & \vdots & -1 \end{bmatrix}$

7. $\begin{bmatrix} 1 & 1 & 1 & \vdots & 5 \\ 2 & 0 & 1 & \vdots & 8 \\ 0 & 3 & 2 & \vdots & 5 \end{bmatrix}$

8. $\begin{bmatrix} -3 & -4 & 5 & \vdots & 4 \\ 2 & 5 & 3 & \vdots & -1 \\ -4 & -1 & 2 & \vdots & -8 \end{bmatrix}$

9. $\begin{bmatrix} 5 & 1 & -1 & \vdots & 9 \\ 1 & -2 & -1 & \vdots & -6 \\ 3 & 4 & 5 & \vdots & 0 \end{bmatrix}$

_____   _____   _____

Solve each system of equations by using the row-reduction method.

10. $\begin{cases} 2x + y = 2 \\ 3x + 2y = 7 \end{cases}$

11. $\begin{cases} 3x + 11y = 10 \\ 2x - 5y = 19 \end{cases}$

12. $\begin{cases} 3x + 4y = 1 \\ 8x + 11y = 4 \end{cases}$

_____   _____   _____

13. $\begin{cases} x + y - 3z = -21 \\ 2x - y + z = 12 \\ 3x + 2y + 2z = 7 \end{cases}$

14. $\begin{cases} 2x + 5y - 3z = -11 \\ 3x - 2y + 4z = 7 \\ 2x + 3y - 2x = -10 \end{cases}$

15. $\begin{cases} 3x + 6y - 4z = -42 \\ 2x + 2y + 3z = 14 \\ 4x + 3y - 5z = -34 \end{cases}$

_____   _____   _____

16. $\begin{cases} x + y + z = 6 \\ 2x - 3y + 5z = -11 \\ x + 3y - 4z = 19 \end{cases}$

17. $\begin{cases} 3x - 2y + z = 16 \\ x + 3y + 4z = 9 \\ 2x - y + 3z = 15 \end{cases}$

18. $\begin{cases} 2x - 4y + 3z = -8 \\ x + 3y - 2z = 9 \\ 3x + 2y + z = 13 \end{cases}$

_____   _____   _____

Practice

5.1 *Introduction to Quadratic Functions*

Show that each function is a quadratic function by writing it in the form $f(x) = ax^2 + bx + c$ and identifying *a*, *b*, and *c*.

1. $f(x) = (x - 3)(x - 5)$ _____

2. $g(x) = (7 - x)(9 - x)$ _____

3. $k(x) = -3(x - 11)(x + 1)$ _____

4. $h(x) = (2x + 5)(3x - 1)$ _____

5. $d(x) = (x - 3)^2 - 4$ _____

Identify whether each function is quadratic. Use a graph to check your answers.

6. $f(x) = -4x + x^2$ _____ 7. $k(x) = \dfrac{1}{x}$ _____

8. $h(x) = \dfrac{2x^3 + x}{x^2 - 1}$ _____ 9. $g(x) = 16 - 3x$ _____

10. $b(x) = x^2 - 2x(x + 1)$ _____ 11. $m(x) = 3x - x(x + 9)$ _____

State whether the parabola opens up or down and whether the *y*-coordinate of the vertex is the minimum value or the maximum value of the function.

12. $f(x) = 5x^2 - 3x$ _____ 13. $g(x) = 4x^2 + 7x - 2$ _____

14. $h(x) = (5 - x)(2 - 3x)$ _____ 15. $q(x) = (4 - x)(2 + 7x)$ _____

Graph each function and give the approximate coordinates of the vertex.

16. $k(x) = 4x^2 - 3$ 17. $h(x) = -x^2 - x + 6$ 18. $p(x) = -(x + 4)(x - 0.5)$

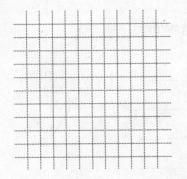 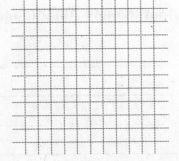 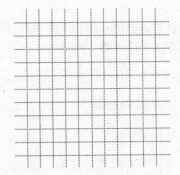

_____ _____

Practice
5.2 *Introduction to Solving Quadratic Equations*

Solve each equation. Give both exact solutions and approximate solutions to the nearest hundredth.

1. $x^2 = 100$

_____

2. $12x^2 = 36$

_____

3. $(x + 3)^2 = 81$

_____

4. $5x^2 - 4 = 96$

_____

5. $x^2 - 12 = 4$

_____

6. $6x^2 + 15 = 23$

_____

7. $4x^2 - 9 = 17$

_____

8. $12 = 4(x - 2)^2 - 8$

_____

9. $14 = 0.5x^2 + 5$

_____

10. $7(x + 1)^2 = 161$

_____

Find the unknown length in each right triangle. Round answers to the nearest tenth.

11.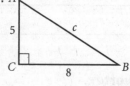

12.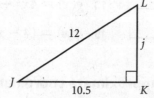

13.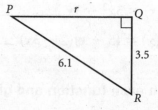

_____   _____

Find the missing side length in right triangle *ABC*. Round answers to the nearest tenth.

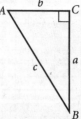

14. $a = 15$ and $b = 7$ _____

15. $a = 2.4$ and $c = 7.3$ _____

16. $b = 2$ and $c = \sqrt{10}$ _____

17. $a = 9.1$ and $b = 7$ _____

Practice

5.3 Factoring Quadratic Expressions

Factor each expression.

1. $12x - 60$ _____

2. $-24x + 4x^2$ _____

3. $(2 - 7x) - 3x(2 - 7x)$ _____

4. $4x(x - 12) - 3(x - 12)$ _____

5. $3x^2 + 21x$ _____

6. $-3x^2 + 105x$ _____

Factor each quadratic expression.

7. $x^2 + 17x + 52$

8. $x^2 - x - 20$

9. $x^2 - 7x - 18$

10. $x^2 + 11x + 28$

11. $x^2 - x - 90$

12. $x^2 + 11x - 26$

13. $4x^2 + 4x + 1$

14. $3x^2 + 5x - 2$

15. $2x^2 - 3x - 2$

Solve each equation by factoring and applying the Zero-Product Property.

16. $x^2 - 256 = 0$

17. $5x^2 - 125 = 0$

18. $x^2 + 9x + 14 = 0$

19. $3x^2 - 6x + 3 = 0$

20. $4x^2 - 12x + 9 = 0$

21. $6x^2 - x - 1 = 0$

22. $9x^2 - 4 = 0$

23. $5x^2 + 1 = 6x$

24. $7x^2 - 2 = 5x$

Use factoring and the Zero-Product Property to find the zeros of each quadratic function.

25. $f(x) = x^2 - x - 12$

26. $g(x) = 2x^2 + 3x - 5$

27. $h(x) = x^2 + 12x - 45$

28. $b(x) = x^2 - 13x + 42$

29. $k(x) = 2x^2 + 3x - 2$

30. $q(x) = 4x^2 + 12x + 9$

Practice
5.4 Completing the Square

Complete the square for each quadratic expression in order to form a perfect-square trinomial. Then write the new expression as a binomial squared.

1. $x^2 + 24x$ _____

2. $x^2 - 40x$ _____

3. $x^2 - 20x$ _____

4. $x^2 + 5x$ _____

5. $x^2 + 9x$ _____

6. $x^2 - 19x$ _____

Solve by completing the square. Round your answers to the nearest tenth, if necessary.

7. $x^2 - 2x - 7 = 0$ 8. $x^2 - 8x + 13 = 0$ 9. $x^2 - 14x - 1 = 0$

_____   _____   _____

10. $x^2 + 20x = 3$ 11. $x^2 + 1 = 5x$ 12. $x^2 - 4 = 6x$

_____   _____   _____

13. $2x^2 - 13 = 2x$ 14. $x^2 + 7x + 2 = 0$ 15. $2x^2 + 16x = 3$

_____   _____   _____

Write each quadratic function in vertex form. Find the coordinates of the vertex and the equation of the axis of symmetry.

16. $f(x) = -\frac{1}{2}x^2$ 17. $f(x) = 7 - 3x^2$

_____   _____

18. $f(x) = x^2 - 12x - 3$ 19. $f(x) = x^2 - 2x - 10$

_____   _____

20. $f(x) = x^2 - 10x - 10$ 21. $f(x) = 3x^2 + 15x - 2$

_____   _____

Practice
5.5 The Quadratic Formula

Use the quadratic formula to solve each equation. Round your answer to the nearest tenth.

1. $x^2 - 10x + 3 = 0$

2. $x^2 - 4x + 1 = 0$

3. $x^2 - 11 = 0$

4. $x^2 + 3x - 15 = 0$

5. $x^2 = 9 - 4x$

6. $x^2 + 7x - 13 = 2x$

7. $14 = 2x^2 + x$

8. $(x - 2)(x - 5) = 2$

9. $2x^2 - 6x = 9$

10. $3x^2 - 6 = 4x$

11. $-2x^2 + 3x + 16 = 0$

12. $4x^2 + x - 1 = 0$

For each quadratic function, find the equation for the axis of symmetry and the coordinates of the vertex. Round your answers to the nearest tenth, if necessary.

13. $y = 2x^2 + 4x - 3$

14. $y = -3x^2 + 9x + 5$

15. $y = 6x^2 - 12x + 5$

16. $y = 3x^2 + 4x - 9$

17. $y = 4x^2 - 8x + 1$

18. $y = 5x^2 + 4x - 1$

Practice
5.6 *Quadratic Equations and Complex Numbers*

Find the discriminant, and determine the number of real solutions. Then solve.

1. $-2x^2 + 5x - 3 = 0$

2. $x^2 + 3x + 9$

3. $3x^2 - 2x + 6 = 0$

4. $4x^2 + 4 = x$

5. $6x^2 - 3x + 4 = 0$

6. $4x = 4x^2 + 7$

Perform the indicated addition or subtraction.

7. $(-6 + 12i) + (4 - i)$

8. $(3 - i) - (-4 + 9i)$

9. $(11 + i) + (2 + 8i)$

10. $(-8 + 4i) + (7 - i)$

11. $(-8 + 4i) - (7 - i)$

12. $(12 + 16i) - (12 + 11i)$

13. $(-7 - 2i) + (3 + 3i)$

14. $(4 - 12i) + 7i$

15. $(-7 + 13i) - (1 + 6i)$

Write the conjugate of each complex number.

16. $15i$ _____

17. $27 - 4i$ _____

18. $-12 + 19i$ _____

Simplify.

19. $4i(-7 + i)$

20. $\dfrac{5 - i}{5 + i}$

21. $(3 - i)(9 + 3i)$

22. $\dfrac{14 - 2i}{3 + i}$

23. $(2 + 5i)^2$

24. $(6 - 3i)(2 + 2i)$

25. $\dfrac{5 - 4i}{i}$

26. $\dfrac{-2 + 3i}{-3 - 2i}$

Practice

5.7 Curve Fitting with Quadratic Models

Solve a system of equations in order to find a quadratic function that fits each set of data points exactly.

1. $(-2, -20), (0, 2), (3, -25)$

   _____

2. $(1, 6), (2, 13), (-2, 21)$

   _____

3. $(4, 9), (6, 21), (-2, -3)$

   _____

4. $(0, -3), (-1, 0), (1, 4)$

   _____

5. $(-2, 29), (2, 17), (1, 2)$

   _____

6. $(3, 0), (-1, -12), (2, 3)$

   _____

7. $(0, -2), (4, -38), (-2, -20)$

   _____

8. $(-3, 1), (-2, -5), (-1, -7)$

   _____

9. $(4, 24), (6, 52), (8, 92)$

   _____

10. $\left(\frac{1}{2}, -2\frac{1}{4}\right), (2, -12), (3, -16)$

   _____

11. $(4, 21), \left(3, 13\frac{1}{2}\right), \left(-1, 3\frac{1}{2}\right)$

   _____

12. $(-2, 11), (-1, -3), (4, 77)$

   _____

A baseball player throws a ball. The table shows the height, y, of the ball x seconds after it is thrown.

Time (seconds)	Height (feet)
0.25	7
0.5	9
1	7

13. Find a quadratic function to model the data. _____

14. What was the maximum height reached by the ball? _____

15. How long did it take the ball to reach its maximum height? _____

16. Use your model to predict the height of the ball 1.25 seconds after it was thrown.   _____

17. Use your model to determine how many seconds it took for the ball to hit the ground.   _____

Practice

5.8 *Solving Quadratic Inequalities*

Solve each inequality. Graph the solution on a number line.
Round irrational numbers to the nearest hundredth.

1. $x^2 - 16 \geq 0$ _____

2. $x^2 + 2x - 8 \leq 0$ _____

3. $x^2 + 7x + 10 \leq 0$ _____

4. $x^2 - 2x + 4 < 0$ _____

5. $x^2 - 3x - 1 > 0$ _____

6. $x^2 + 10x - 3 < 7x$ _____

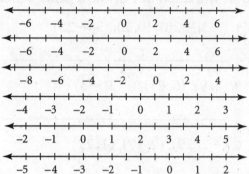

Graph each inequality and shade the solution region.

7. $y > x^2 + \frac{1}{2}x$

8. $y \leq (x + 3)^2 - 2$

9. $y > -(x - 1)^2 + 1$

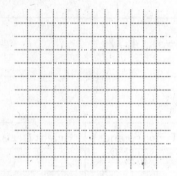
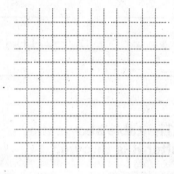
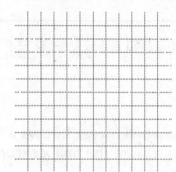

10. $y \geq x^2 - 4x - 5$

11. $y \leq -(x - 3)^2 + 3$

12. $y < 2x^2 - x - 1$

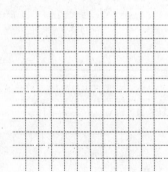
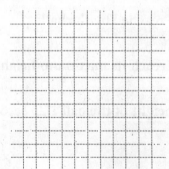
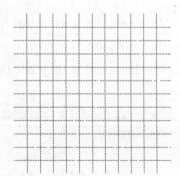

Practice

6.1 Exponential Growth and Decay

Find the multiplier for each rate of exponential growth or decay.

1. 1% growth _____
2. 1% decay _____
3. 7% decay _____
4. 12% growth _____
5. 10% growth _____
6. 3% decay _____
7. 5.2% decay _____
8. 7.5% growth _____
9. 0.4% growth _____
10. 5.9% decay _____

Evaluate each expression to the nearest thousandth for the given value of x.

11. 2^x for $x = 0.5$ _____
12. $10(2^x)$ for $x = \frac{2}{3}$ _____
13. $\left(\frac{1}{2}\right)^{3x}$ for $x = 2$ _____
14. $7(0.5)^x$ for $x = -2$ _____
15. $42 \cdot 2^{x-1}$ for $x = \frac{3}{4}$ _____
16. $20 \cdot 2^{2x}$ for $x = 1.75$ _____
17. $15\left(\frac{1}{2}\right)^{2x-1}$ for $x = 2$ _____
18. $66\left(\frac{1}{2}\right)^x$ for $x = 3$ _____
19. $512(2)^{3x}$ for $x = 0.1$ _____
20. $12(2)^{x-2}$ for $x = 6.5$ _____

Predict the result in each situation.

21. The population of a city in 1990 was 1,215,112. The population was growing at a rate of about 5% per decade. Predict the population of the city

    a. in the year 2000.
    b. in the year 2005.

22. The initial population of bacteria in a lab test is 400. The number of bacteria doubles every 30 minutes. Predict the bacteria population at the end of

    a. two hours.
    b. three hours.

Practice

6.2 *Exponential Functions*

Identify each function as linear, quadratic or exponential.

1. $f(x) = (x + 1)^2 - x$ 2. $g(x) = 5x - 4^2$ 3. $k(x) = 2x + 11$

_____   _____   _____

4. $g(x) = 2^x + 11$ 5. $w(x) = x^2 + 11$ 6. $h(x) = 0.4^{2x}$

_____   _____   _____

7. $b(x) = x(x - 4) + (4 - x^2)$ 8. $f(x) = \left(\frac{2}{3}\right)^{3x}$ 9. $h(x) = 450(0.3)^{-x}$

_____   _____   _____

Tell whether each function represents exponential growth or decay.

10. $f(x) = 5.9(2.6)^x$ 11. $b(x) = 13(0.7)^x$ 12. $k(x) = 22(0.15)^x$

_____   _____   _____

13. $m(x) = 51(4.3)^x$ 14. $w(x) = 0.72 \cdot 2^x$ 15. $z(x) = 47(0.55)^x$

_____   _____   _____

16. $h(x) = 2.5(0.8)^x$ 17. $g(x) = 0.8(3.2)^x$ 18. $a(x) = 150(1.1)^x$

_____   _____   _____

Find the final amount for each investment.

19. $1300 earning 5% interest compounded annually for 10 years _____

20. $850 earning 4% interest compounded annually for 6 years _____

21. $720 earning 6.2% interest compounded semiannually for 5 years _____

22. $1100 earning 5.5% interest compounded semiannually for 2 years _____

23. $300 earning 4.5% interest compounded quarterly for 3 years _____

24. $1000 earning 6.5% interest compounded quarterly for 4 years _____

25. $5000 earning 6.3% interest compounded daily for 1 year _____

26. $2000 earning 5.5% interest compounded daily for 3 years _____

Practice

6.3 *Logarithmic Functions*

Write each equation in logarithmic form.

1. $19^2 = 361$

2. $20^3 = 8000$

3. $3375^{\frac{1}{3}} = 15$

4. $\left(\frac{3}{4}\right)^{-3} = 64$

5. $\left(\frac{3}{7}\right)^3 = \frac{27}{343}$

6. $11^{-3} = \frac{1}{1331}$

Write each equation in exponential form.

7. $\log_{12} 144 = 2$

8. $\log_5 15,625 = 6$

9. $\log_{21} 9261 = 3$

10. $\log_{3600} 60 = \frac{1}{2}$

11. $\log_{11} \frac{1}{14,641} = -4$

12. $\log_{\frac{1}{5}} 625 = -4$

Solve each equation for x. Round your answers to the nearest hundredth.

13. $10^x = 35$

14. $10^x = 91$

15. $10^x = 0.2$

16. $10^x = 1.8$

17. $10^x = 0.08$

18. $10^x = 1055$

Find the value of v in each equation.

19. $v = \log_{10} 1000$

20. $v = \log_{15} 225$

21. $v = \log_{12} 144$

22. $8 = \log_2 v$

23. $-4 = \log_5 v$

24. $-3 = \log_7 v$

25. $-2 = \log_v \frac{1}{100}$

26. $\log_v 729 = 6$

27. $\log_v \frac{1}{256} = -4$

Practice

6.4 Properties of Logarithmic Functions

Write each expression as a sum or a difference of logarithms. Then simplify, if possible.

1. $\log_{10}(4 \cdot 100)$

_____

2. $\log_5 \frac{72}{25}$

_____

3. $\log_7(5 \cdot 3 \cdot 4)$

_____

4. $\log_3 15q$

_____

5. $\log_8 \frac{64}{y}$

_____

6. $\log_9 \frac{3a}{7}$

_____

Write each expression as a single logarithm. Then simplify, if possible.

7. $\log_3 5 + \log_3 6$

_____

8. $\log_5 x - \log_5 2$

_____

9. $\log_8 2 + \log_8 32$

_____

10. $\log_9 5 + \log_9 y - \log_9 4$

_____

11. $2\log_{12} 6 + \log_{12} 4$

_____

12. $\frac{1}{2}\log_3 81 + \log_3 15$

_____

13. $\log_b m + \log_b 2 - \log_b x$

_____

14. $3\log_b x - (\log_b 4 + \log_b x)$

_____

15. $3\log_b z + \log_b y - 4\log_b z$

_____

Evaluate each expression.

16. $5^{\log_5 12}$ _____

17. $12^{\log_{12} 73}$ _____

18. $\log_3 3^{2.5}$ _____

19. $\log_2 2^{4.7}$ _____

20. $\log_4 4^3 - \log_3 81$ _____

21. $9^{\log_9 15} - \log_3 3^5$ _____

Solve for x and check your answers.

22. $\log_2(10x) = \log_2(3x + 14)$ _____

23. $2\log_3 x = \log_3 4$ _____

24. $\log_5(4x - 3) = \log_5(x + 1)$ _____

25. $\log_7(x^2 - 1) = \log_7 8$ _____

26. $\log_8(x^2 - 3x) = \log_8(2x + 6)$ _____

27. $2\log_2(x + 2) = \log_2(3x + 16)$ _____

28. $\log_b 8 = \log_b x + \log_b(x - 2)$ _____

29. $2\log_b(x + 1) = \log_b(-x + 11)$ _____

Practice

6.5 Applications of Common Logarithms

Solve each equation. Round your answers to the nearest hundredth.

1. $5^x = 16$

2. $6^x = 5.5$

3. $2^x = 100$

4. $8^x = 12$

5. $3^x = 22$

6. $9^x = 0.35$

7. $5.5^x = 6$

8. $7^x = 0.8$

9. $3^{-x} = 0.2$

10. $12^x = 18$

11. $4.22^{2x} = 61$

12. $8.2^{x+1} = 55$

13. $\left(\frac{1}{2}\right)^{-x} = 17$

14. $14^x = 33.8$

15. $72^{2x} = 35$

Evaluate each logarithmic expression to the nearest hundredth.

16. $\log_7 30.6$

17. $\log_3 11$

18. $\log_2 13$

19. $\log_5 0.4$

20. $\log_4 83$

21. $\log_9 2.4$

22. $\log_6 8$

23. $\log_2 8.5$

24. $\log_4 6.1$

25. $\log_3 0.6$

26. $\log_8 0.32$

27. $\log_5 10$

28. $3 - \log_{\frac{1}{2}} 20$

29. $\log_{\frac{1}{4}} 9 + 5$

30. $1 - \log_7 25$

Practice

6.6 The Natural Base, e

Evaluate each expression to the nearest thousandth.

1. e^8

2. $e^{2.5}$

3. $e^{5.2}$

4. $2e^4$

5. $\ln 35$

6. $\ln 12.6$

7. $\ln(-1.4)$

8. $\ln \sqrt{12}$

9. $\ln 112$

Write an equivalent exponential or logarithmic equation.

10. $e^x \approx 55$

11. $\ln 44 \approx 3.78$

12. $e^{-3} \approx 0.05$

13. $\ln 10 \approx 2.30$

14. $e^4 \approx 54.6$

15. $\ln 125 \approx 4.83$

16. $e^5 \approx 148$

17. $\ln 1 = 0$

18. $e^{-0.8} \approx 0.45$

Solve each equation for *x* by using the natural logarithm function.
Round your answers to the nearest hundredth.

19. $33^x = 74$

20. $15^x = 19.5$

21. $4.8^x = 30$

22. $0.7^x = 22$

23. $1.5^x = 70$

24. $4^{\frac{2}{3}x} = 0.5$

25. $15^{-x} = 24$

26. $0.25^{2x} = 41$

27. $44^x = 19$

28. $1000 is deposited in an account with an interest rate of 6.5%.
Interest is compounded continuously, and no deposits or withdrawals
are made. Find the amount in the account at the end of three years. _____

Practice

6.7 *Solving Equations and Modeling*

Solve each equation for *x*. Write the exact solution and the approximate solution to the nearest hundredth, when appropriate.

1. $7^x = 7^4$

2. $5^{3x} = 25$

3. $\log_4 x = \frac{1}{2}$

4. $\log x = 3$

5. $5 = \log_x \frac{1}{32}$

6. $9^x = 6$

7. $e^{3x} = 15$

8. $\ln(2x - 7) = \ln 3$

9. $10^x + 4 = 32$

10. $\log_3(2x + 1) = 2$

11. $3 \ln x = \ln 16 + \ln 4$

12. $\ln 2x - \ln(x - 2) = \ln 3$

13. $\log_x \frac{1}{9} = -2$

14. $\log_{\frac{1}{8}} \frac{1}{16} = x$

15. $\ln 2x = 3$

16. $5\left(1 + e^{\frac{x}{3}}\right) = 8.2$

In Exercises 17 and 18, use the equation $M = \frac{2}{3} \log \frac{E}{10^{11.8}}$, where M represents the magnitude and E represents the energy released.

17. On January 17, 1994, an earthquake with a magnitude of 6.6 injured more than 8000 people and caused an estimated \$13–20 billion of damage to the San Fernando Valley in California. Find the amount of energy released by the earthquake. _____

18. On January 17, 1995, an earthquake struck Osaka, Kyoto, and Kobe, Japan, injuring more than 36,000 people and causing an estimated \$100 billion of damage. The quake released about 3.98×10^{22} ergs of energy. Find the earthquake's magnitude on the Richter scale. Round your answer to the nearest tenth. _____

Practice

7.1 An Introduction to Polynomials

Determine whether each expression is a polynomial. If so, classify the polynomial by degree and by number of terms.

1. $5x^2 - 22x^5 + 17x$

2. $\frac{x}{2} - \frac{x^2}{2} + 13$

3. $\frac{7}{x^2} + \frac{13}{x^3}$

_____

_____

_____

4. $-2x^3 - 4x^2 - 15x + 7$

5. $\sqrt[3]{x} + 12\sqrt{x}$

6. $43x^{-6} + 9x^{-7} + 12x^{-1}$

_____

_____

_____

Evaluate each polynomial expression for the indicated value of x.

7. $2x^3 - 3x^2 + 4x, x = -2$ _____

8. $-x^4 + 3x^3 - 2x^2 + 4, x = -1$ _____

9. $x^5 + x^4 + x^3 + x^2 + 1, x = 2$ _____

10. $0.5x^3 - 0.6x^2 - 3x, x = 10$ _____

11. $7x^2 - 19x, x = 5$ _____

12. $0.75x^3 - 15x^2 + 10x, x = 4$ _____

Write each sum or difference as a polynomial in standard form.

13. $(3x^4 + 12x^3 - 2x^2) + (5x^4 - x^3 + 7x^2)$

14. $(-7x^4 + 24x^5 - 3x^2 + 9) - (2x^5 + 6x^4 + x + 1$

_____

_____

15. $(8.8x + 2 + 3x^2 - x^4) - (5x^3 + 10x - 7x^2)$

16. $(7.1x^3 + 3.2x^2 - 7x + 8) + (9x^2 - 2x^3 + 18)$

_____

_____

Sketch the graph of each function. Describe the general shape of the graph.

17. $a(x) = -2x^4 + 5x^3 - 2$

18. $k(x) = 4x^4 + 4x^3 - 6x^2$

19. $f(x) = \frac{3}{4}x^3 + 2x^2 + 1$

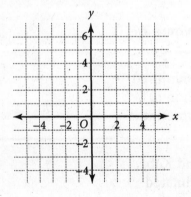

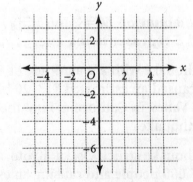

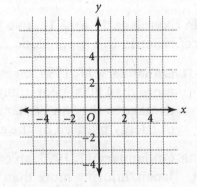

_____

_____

_____

Algebra 2

Practice

7.2 *Polynomial Functions and Their Graphs*

Graph each function and approximate any local maxima or minima to the nearest tenth.

1. $P(x) = x^2 + 3x + 4$

2. $P(x) = 6 + x - 3x^2$

_____

_____

3. $P(x) = 2x^3 - 2x^2 + 1$

4. $P(x) = x^4 + x^3 - 4x^2 - 2x + 2$

_____

_____

Graph each function. Find any local maxima or minima to the nearest tenth. Find the intervals over which the function is increasing and decreasing.

5. $P(x) = 4x^3 - 3x^2 + 2, -6 \le x \le 6$ _____

_____

6. $P(x) = 0.3x^4 + x^3 - x, -4 \le x \le 4$ _____

_____

7. $P(x) = x^3 + 1.2x^2 - 2, -5 \le x \le 5$ _____

_____

8. $P(x) = -x^4 + 2.5x^3 - x^2 + 1, -4 \le x \le 4$ _____

_____

Describe the end behavior of each function.

9. $12 - 4.2x^3 + x^2$

10. $3.3x^3 - 2x^2 - 5x + 1$

_____

_____

11. $5x^3 - 6x^4 + x^2 + 1$

12. $1.1x^4 - 2.2x^3 + 3.3x^2 - 4$

_____

_____

13. Factory sales of passenger cars, in thousands, in the United States are shown in the table below. Find a quartic regression model for the data by using $x = 0$ for 1990. (*Source: Bureau of the Census*)

1990	1991	1992	1993	1994	1995
6050	5407	5685	5969	6549	6310

Practice

7.3 Products and Factors of Polynomials

Write each product as a polynomial in standard form.

1. $0.5x(16x^4 - 10x^3 + 6x)$

2. $(x - 10)(2x + 3)$

3. $(x - 4)(5x^2 + 3x + 7)$

_____

4. $(x + 2)(x - 8)(x - 1)$

5. $(2x - 5)(x + 1)^2$

6. $(3x - 1)^3$

_____

Use substitution to determine whether the given linear expression is a factor of the given polynomial.

7. $x^2 + 2x - 12; x + 4$ _____

8. $3x^2 - x - 4; x + 1$ _____

9. $x^3 - 9x + 1; x - 3$ _____

10. $2x^3 - 11x^2 + 8x - 15; x - 5$ _____

11. $2x^3 + 10x^2 - 28x; x + 7$ _____

12. $3x^3 - 2x^2 - 6x - 2; x - 2$ _____

Divide by using long division.

13. $(2x^2 + 7x - 30) \div (x + 6)$

14. $(6x^2 + 2x - 5) \div (3x + 5)$

_____

15. $(8x^3 + 12x^2 + 6x + 5) \div (2x + 1)$

16. $(5x^3 + x^2 - x + 3) \div (x + 1)$

_____

Divide by using synthetic division.

17. $(x^3 - x^2 + x - 21) \div (x^2 + 2x + 7)$

18. $(x^3 - 5x^2 - 20x - 32) \div (x - 8)$

_____

19. $(x^3 + 4x^2 + 4x + 3) \div (x^2 + x + 1)$

20. $(x^5 - x^3 - 3) \div (x^2 - 3)$

_____

For each function below, use synthetic division and substitution to find the indicated value.

21. $P(x) = x^2 + 3x + 1; P(2)$ _____

22. $P(x) = x^3 - 2x + 4; P(3)$ _____

23. $P(x) = 2x^4 - 3x^3 + 2x^2 - 6; P(2)$ _____

24. $P(x) = 3x^4 - 4x^3 + x^2 - 1; P(-2)$ _____

Algebra 2

Practice

7.4 Solving Polynomial Equations

Use factoring to solve each equation.

1. $x^3 - 81x = 0$

2. $x^3 - 11x^2 + 10x = 0$

3. $2x^3 - x^2 - x = 0$

_____   _____   _____

4. $x^3 + 2x^2 = 15x$

5. $2x^3 - 2x^2 - 24x = 0$

6. $3x^3 + x = 4x^2$

_____   _____   _____

Use graphing, synthetic division, and factoring to find all of the roots of each equation.

7. $x^3 - 3x^2 - 4x + 12 = 0$

8. $x^3 + 4x^2 + x = 6$

9. $3x^3 + 2x^2 - 37x + 12$

_____   _____   _____

10. $x^3 + 29x + 42 = 12x^2$

11. $x^3 - 11x^2 + 24x + 36 = 0$

12. $x^3 + 64 = 4x^2 + 16x$

_____   _____   _____

Use variable substitution and factoring to find all of the roots of each equation. If necessary, leave your answers in radical form.

13. $x^4 - 10x^2 + 24 = 0$

14. $x^4 - 10x^2 + 21 = 0$

15. $x^4 + 54 = 15x^2$

_____   _____   _____

16. $x^4 - 7x^2 = -10$

17. $x^4 - 17x^2 + 16 = 0$

18. $x^4 + 20 = 12x^2$

_____   _____   _____

Use a graph and the Location Principle to find the real zeros of each function. Give approximate values to the nearest tenth, if necessary.

19. $P(x) = 2x^3 - 4x + 1$

20. $P(x) = 1.5x^3 + 2x^2 - 0.25$

_____   _____

21. $P(x) = 2.5x^4 - 2x^2$

22. $P(x) = 12x^3 - 15x^2 + x + 1$

_____   _____

23. $P(x) = 8x^3 - 6x^2 - 2x + 1$

24. $P(x) = 0.5x^4 + 2x^3 - 5x - 1$

_____   _____

Practice

7.5 Zeros of Polynomial Functions

Find all the rational roots of each polynomial equation.

1. $P(x) = 5x^2 - 6x + 1$

2. $P(x) = 3x^3 - x^2 - 12x + 4$

3. $P(x) = 6x^3 + x^2 - 4x + 1$

4. $P(x) = 4x^3 - 11x^2 + 5x + 2$

5. $P(x) = 2x^4 - 5x^3 - 12x^2 - x + 4$

6. $P(x) = 18x^4 - 9x^3 - 17x^2 + 4x + 4$

Find all the zeros of each polynomial function.

7. $P(x) = x^3 + 5x^2 + 2x + 10$

8. $P(x) = x^3 + 3x^2 + 4x + 12$

9. $P(x) = x^3 - 3x^2 + 12x - 36$

10. $P(x) = x^3 - 5x^2 + 9x - 45$

11. $P(x) = x^3 + 2x^2 - 16$

12. $P(x) = x^4 - 5x^3 + 5x^2 - 5x + 4$

Find all real values of x for which the functions are equal. Round your answers to the nearest hundredth.

13. $P(x) = x^4 - 5$, $Q(x) = x + 1$

14. $P(x) = x^3 - 5x + 3$, $Q(x) = -x^3 + 5x + 3$

15. $P(x) = x^3 + 2x + 1$, $Q(x) = 4x + 1$

16. $P(x) = x^4 + x^3 + 2x - 1$, $Q(x) = x^2 + 2x + 1$

Write a polynomial function, P, in standard form by using the given information.

17. The zeros of $P(x)$ are -3, 2, and 4, and $P(0) = 120$. _____

18. $P(0) = -96$, and two of the three zeros are -3 and $4i$. _____

Practice

8.1 Inverse, Joint, and Combined Variation

For Exercises 1–4, *y* varies inversely as *x*. Write the appropriate inverse-variation equation, and find *y* for the given values of *x*.

1. $y = 12$ when $x = 7$; $x = 5, 10, 16,$ and 20

_____

2. $y = 0.4$ when $x = 2$; $x = 0.1, 5, 8,$ and 20

_____

3. $y = 3\frac{1}{3}$ when $x = 45$; $x = 6, 15, 20,$ and 60

_____

4. $y = 12$ when $x = 0.4$; $x = 0.5, 6, 10,$ and 16

_____

For Exercises 5–8, *y* varies jointly as *x* and *z*. Write the appropriate joint-variation equation, and find *y* for the given values of *x* and *z*.

5. $y = 16$ when $x = 4$ and $z = 0.5$;
   $x = 2$ and $z = 0.25$

_____

6. $y = 120$ when $x = 2.5$ and $z = 2$;
   $x = 3$ and $z = 2$

_____

7. $y = 12$ when $x = 4$ and $z = 5$;
   $x = 6$ and $z = 3$

_____

8. $y = 192$ when $x = 2$ and $z = 3$
   $x = 0.6$ and $z = 5$

_____

For Exercises 9–12, *z* varies jointly as *x* and *y* and inversely as *w*. Write the appropriate combined-variation equation, and find *z* for the given values of *x*, *y*, and *w*.

9. $z = 320$ when $x = 4, y = 10,$ and $w = 2.5$;
   $x = 5, y = 6,$ and $w = 8$

_____

10. $z = 3.2$ when $x = 0.2, y = 8,$ and $w = 4$;
    $x = 3, y = 6,$ and $w = 16$

_____

11. $z = 3.75$ when $x = 6, y = 12,$ and $w = 48$;
    $x = 0.05, y = 40,$ and $w = 0.5$

_____

12. $z = 4.8$ when $x = 0.2, y = 10,$ and $w = 5$;
    $x = 2\frac{1}{3}, y = 5,$ and $w = 8$

_____

13. The *apothem* of a regular polygon is the perpendicular distance from the center of the polygon to a side. The area, *A*, of a regular polygon varies jointly as the apothem, *a*, and the perimeter, *p*. A regular triangle with an apothem of 3 inches and a perimeter of 31.2 inches has an area of 46.8 square inches. Find the constant of variation and write a joint-variation equation. Then find the area of a regular triangle with an apothem of 2.3 inches and a perimeter of 12 inches.

_____

_____

_____

Practice

8.2 Rational Functions and Their Graphs

Determine whether each function below is a rational function. If so, find the domain. If the function is not rational, state why not.

1. $f(x) = \dfrac{x^3 - 5x + 7}{x^2 - 3}$

2. $h(x) = \dfrac{x + 2}{|x| - 2}$

3. $w(x) = \dfrac{12 - 2x}{x^2 - 1}$

_____

Identify all vertical and horizontal asymptotes of the graph of each rational function.

4. $k(x) = \dfrac{2x + 1}{x - 9}$ _____

5. $p(x) = \dfrac{2x^2 + 3}{(x - 1.5)^2}$ _____

6. $m(x) = \dfrac{3x - 8}{x^2 - 7}$ _____

Find the domain of each rational function. Identify all asymptotes and holes in the graph of each rational function.

7. $h(x) = \dfrac{4x - 3}{x^2 - 6x}$ _____

8. $g(x) = \dfrac{x - 1}{x^2 + 4x - 5}$ _____

9. $n(x) = \dfrac{3x^2 + 12x}{x^2 + 7x + 12}$ _____

Sketch the graph of each rational function. Identify all asymptotes and holes in the graph of the function.

10. $a(x) = \dfrac{3x}{x - 4}$

11. $f(x) = \dfrac{x + 2}{2x^2 + 3x - 2}$

12. $b(x) = \dfrac{x + x^2}{x^2 - 1}$

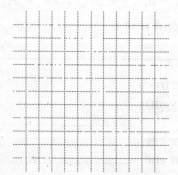
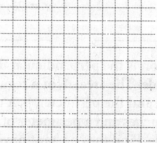
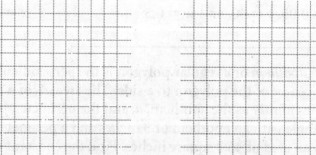

Algebra 2

Practice

8.3 Multiplying and Dividing Rational Expressions

Simplify each rational expression.

1. $\dfrac{2x^4}{x^5} \cdot \dfrac{6x}{x^3} \cdot \dfrac{x}{4}$

2. $\dfrac{x^2 - 8x + 7}{x^2 + 6x - 7}$

3. $\dfrac{9x^2 + 12x + 4}{9x^2 - 4}$

4. $\dfrac{3x}{x^{10}} \cdot \dfrac{x^3}{27} \cdot \dfrac{9x^4}{2}$

5. $\dfrac{x^2 + 7x + 12}{x^2 + x - 6}$

6. $\dfrac{4x^4}{9x} \cdot \dfrac{9x^3}{10x} \cdot \dfrac{15x^2}{2x}$

7. $\dfrac{x^2 - 5x + 6}{x + 4} \cdot \dfrac{3x + 12}{x - 2}$

8. $\dfrac{2x - 3}{5x + 1} \div \dfrac{6x^2 - 13x + 6}{15x^2 - 7x - 2}$

9. $\dfrac{4x - 8}{x^2 - x - 6} \div \dfrac{x^3 + x^2 - 6x}{x^2 - 9}$

10. $\dfrac{x^3 - 9x}{x^2 + 11x + 24} \cdot \dfrac{x^2 + 7x - 8}{x^2 - 4x + 3}$

11. $\dfrac{\dfrac{x^2 - 16}{x - 3}}{\dfrac{x + 4}{x^2 - 9}}$

12. $\dfrac{\dfrac{x - 5}{x^2 - 100}}{\dfrac{x^2 - 25}{x + 10}}$

13. $\dfrac{\dfrac{x^2 + 10x - 11}{x^2 + 6x + 5}}{\dfrac{x^2 + 9x - 22}{x^2 + 3x - 10}}$

14. $\dfrac{\dfrac{x^4 - 81}{3x^2 + 27}}{\dfrac{x^2 - x - 12}{x}}$

15. $\dfrac{x - 6}{x + 2} \cdot \dfrac{\dfrac{2x - 1}{x - 6}}{\dfrac{x - 2}{x + 2}}$

16. $\dfrac{\dfrac{x^2 - 5x + 6}{x^2 - 8x + 15}}{\dfrac{x - 2}{x - 5}} \div \dfrac{x^2 - 9}{x^2 + 3x}$

17. $\dfrac{x - 4}{x - 7} \cdot \dfrac{\dfrac{x^2 - 49}{3x - 12}}{\dfrac{x^2 + 14x + 49}{x + 5}}$

18. $\dfrac{\dfrac{x^2 - y^2}{5x^3 y^2}}{\dfrac{4x + 4y}{15x^2 y^5}}$

NAME _____ CLASS _____ DATE _____

Practice
8.4 Adding and Subtracting Rational Expressions

Simplify.

1. $\dfrac{x-7}{3}+\dfrac{x+2}{4}$

2. $\dfrac{5x-4}{x^3+1}-\dfrac{2x+3}{x^3+1}$

3. $\dfrac{3x+4}{3x}-\dfrac{2x+1}{2x}$

4. $\dfrac{x-2}{x^2-4}+\dfrac{2}{3x-6}$

5. $\dfrac{x-2}{x+3}+\dfrac{x+3}{x-2}$

6. $\dfrac{4x}{x^2-16}-\dfrac{4}{x+4}$

7. $\dfrac{x+4}{2x^2-2x}-\dfrac{5}{2x-2}$

8. $\dfrac{x-2}{x+8}-\dfrac{x-2}{x^2+6x-16}$

9. $\dfrac{x+5}{x^2+10x+25}-\dfrac{2x}{x^2-25}$

10. $\dfrac{\frac{x+14}{x-3}}{2}$

11. $\dfrac{\frac{12}{x-2}}{\frac{3}{x-2}}+\dfrac{7}{x-2}$

12. $\dfrac{\frac{x}{x-6}}{\frac{2x-1}{x-6}}-\dfrac{3}{2x-1}$

13. $\dfrac{\frac{10}{3x+1}}{\frac{5x}{3x+1}}+\dfrac{x^2+1}{x}$

14. $\dfrac{\frac{x}{x-2}}{\frac{2x^2}{2-x}}+\dfrac{1}{x+1}$

15. $\dfrac{\frac{x-3}{2x+1}}{\frac{2x-1}{x+3}}-\dfrac{x^2-9}{4x^2-1}$

Write each expression as a single rational expression in simplest form.

16. $\dfrac{5x-2}{x^2-49}+\dfrac{x-15}{x^2-49}-\dfrac{3x+4}{x^2-49}$

17. $\dfrac{2x+3}{x^2-9}-\dfrac{2x-3}{x^2-9}+\dfrac{1}{x+3}$

18. $\dfrac{x}{x-5}-\dfrac{x^2+25}{25-x^2}+\dfrac{5}{x+5}$

19. $\dfrac{x+1}{x-2}+\dfrac{x+2}{x-4}+\dfrac{16-5x}{x^2-6x+8}$

20. $\dfrac{5x}{x^2-9}-\dfrac{4}{x+3}+\dfrac{2}{3-x}$

21. $\dfrac{-2x^2-5x}{x^2+7x}+\dfrac{x-2}{x+7}+\dfrac{2x-3}{x}$

22. $\dfrac{2x-3}{3x^2-13x-10}+\dfrac{2x+1}{5-x}+\dfrac{1}{3x+2}$

23. $\dfrac{5}{xy+3y-2x-6}+\dfrac{4}{x+3}-\dfrac{2}{2-y}$

Algebra 2

Practice

8.5 Solving Rational Equations and Inequalities

Solve each equation. Check your solution.

1. $\dfrac{2x+1}{4x-4} = \dfrac{4}{5}$

2. $\dfrac{x-5}{x-8} = \dfrac{x+1}{x-5}$

3. $\dfrac{x-15}{x+5} = \dfrac{x-12}{x}$

_____

4. $\dfrac{x-8}{x+5} = \dfrac{x-1}{2x+10}$

5. $\dfrac{x^2+1}{x+2} = 3x-1$

6. $\dfrac{x-10}{2x+1} = \dfrac{4x}{3x+4}$

_____

7. $\dfrac{x-2}{x} - 1 = \dfrac{2x+3}{x}$

8. $\dfrac{3}{4} - \dfrac{1}{x} = \dfrac{1}{2x}$

9. $\dfrac{x}{x-2} - \dfrac{x-5}{5} = \dfrac{x-2}{5}$

_____

10. $\dfrac{3}{x-1} + 4 = \dfrac{1}{1-x^2}$

11. $\dfrac{x+3}{x-2} - \dfrac{14}{x+2} = \dfrac{3x-2}{x^2-4}$

12. $\dfrac{3}{x-2} + \dfrac{5}{x+2} = \dfrac{4x^2}{x^2-4}$

_____

Solve each inequality. Check your solution.

13. $\dfrac{x}{x-2} < 2$

14. $\dfrac{x}{x-6} < 2$

15. $\dfrac{x}{x+1} < \dfrac{x}{x-1}$

_____

16. $\dfrac{x}{x-3} > \dfrac{4}{x-2}$

17. $\dfrac{x+1}{x-1} > 2$

18. $\dfrac{x+1}{x+2} - \dfrac{x}{x+3} \le \dfrac{7}{x^2+5x+6}$

_____

19. $\dfrac{x}{x+1} - \dfrac{2}{x-1} > 1$

20. $\dfrac{x}{x+3} + \dfrac{1}{x-4} < 1$

21. $\dfrac{x+1}{x+1} - \dfrac{x}{x+1} > \dfrac{2}{x^2-1}$

_____

Use a graphics calculator to solve each rational inequality.
Round answers to the nearest tenth.

22. $\dfrac{2}{x-3} \le x+3$

23. $\dfrac{x+2}{x+4} < 1-x$

24. $\dfrac{x-3}{x-4} > x$

_____

25. $\dfrac{x+1}{x-2} < \dfrac{1}{x-3}$

26. $\dfrac{x-4}{x} - \dfrac{x}{x-4} < 1$

27. $\dfrac{2x-3}{x} - \dfrac{3}{x-2} > 5$

_____

Practice

8.6 Radical Expressions and Radical Functions

Find the domain of each radical function.

1. $f(x) = \sqrt{12x - 30}$ 2. $f(x) = \sqrt{7(x - 4)}$ 3. $f(x) = \sqrt{x^2 - 36}$

_____   _____   _____

4. $f(x) = \sqrt{4x^2 - 25}$ 5. $f(x) = \sqrt{x^2 - 10x + 25}$ 6. $f(x) = \sqrt{x^2 + 4x + 3}$

_____   _____   _____

Find the inverse of each quadratic function. Then graph the function and its inverse in the same coordinate plane.

7. $y = x^2 - 6x + 8$ 8. $y = 2 - x^2$ 9. $y = x^2 - 2x - 5$

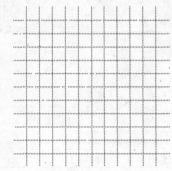   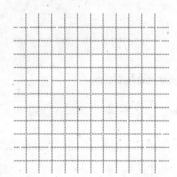 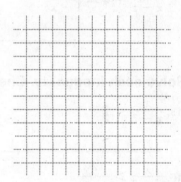

_____   _____   _____

Evaluate each expression. Give exact answers.

10. $\sqrt[3]{\dfrac{192}{3}}$ 11. $\dfrac{3}{4}\sqrt[4]{10{,}000}$ 12. $15\sqrt[3]{-\dfrac{8}{125}}$

_____   _____   _____

13. $4\sqrt[3]{-216}$ 14. $-8\sqrt[3]{-\dfrac{1}{8}}$ 15. $\dfrac{2}{3}\sqrt[3]{-27}$

_____   _____   _____

16. The volume of a sphere with diameter d is given by the equation $V = \frac{1}{6}\pi d^3$. Solve this equation for d in terms of V. Then use your equation to find the diameter, to the nearest foot, of a sphere with a volume of 1000 cubic feet.

_____

Practice

8.7 *Simplifying Radical Expressions*

Simplify each radical expression by using the Properties of *n*th Roots.

1. $\sqrt[5]{32}$

2. $\sqrt[4]{81}$

3. $\sqrt{288\,x^2y^4}$

4. $\sqrt[3]{343x^5y^9z}$

5. $(80x^5)^{\frac{1}{2}}$

6. $(-16x^3y^4)^{\frac{1}{3}}$

Simplify each product or quotient. Assume that the value of each variable is positive.

7. $\sqrt[5]{16x^6} \cdot \sqrt[5]{2x^4}$

8. $\sqrt[3]{16x^2y^5} \cdot \sqrt[3]{4x^2y}$

9. $\sqrt{2x^3y} \cdot \sqrt{5x^3y^3} \cdot \sqrt{10x^2y}$

10. $\dfrac{\sqrt[5]{64x^3y^7z^3}}{\sqrt[5]{2xy}}$

11. $\dfrac{\sqrt[3]{96x^2y^5z^4}}{\sqrt[3]{4yx}}$

12. $\dfrac{\sqrt[4]{64x^{10}y^{10}}}{\sqrt[4]{2x}}$

Find each sum, difference, or product. Give your answer in simplest radical form.

13. $\left(16 + 3\sqrt{2}\right) + \left(9 - \sqrt{2}\right)$

14. $\left(5 + 7\sqrt{3}\right) - \left(2 - 3\sqrt{12}\right)$

15. $\left(11 - 3\sqrt{18}\right) - \left(6 - 4\sqrt{8}\right)$

16. $\left(4 + 3\sqrt{2}\right)\left(3 - 6\sqrt{2}\right)$

17. $\left(4 - 7\sqrt{5}\right)\left(3 + 2\sqrt{5}\right)$

18. $8\sqrt{2}\left(\sqrt{8} - 3\sqrt{2} + 7\sqrt{32}\right)$

Write each expression with a rational denominator and in simplest form.

19. $\dfrac{4}{\sqrt{8}}$

20. $\dfrac{\sqrt{64}}{\sqrt{2}}$

21. $\dfrac{7}{\sqrt{2} + 1}$

22. $\dfrac{8}{2 - \sqrt{3}}$

23. $\dfrac{5}{\sqrt{2} + \sqrt{3}}$

24. $\dfrac{9}{\sqrt{7} - \sqrt{2}}$

Practice

8.8 Solving Radical Equations and Inequalities

Solve each radical equation by using algebra. If the equation has no solution, write *no solution*. Check your solution.

1. $\sqrt{x+5} = 10$

2. $\sqrt{x-6} = 2$

3. $\sqrt{x^2-4} = 2\sqrt{3}$

4. $\sqrt{x+2} = \sqrt{x}$

5. $\sqrt{2x+3} = x+1$

6. $2\sqrt{x-2} = x-2$

7. $3\sqrt{2x+3} = \sqrt{x-7}$

8. $\sqrt{3x-5} = x-1$

9. $\sqrt[3]{x+4} = \sqrt[3]{3x-6}$

Solve each radical inequality by using algebra. If the inequality has no solution, write *no solution*. Check your solution.

10. $\sqrt{x-3} \leq 3$

11. $\sqrt{x+2} > 4$

12. $\sqrt{3x} < 5$

13. $\sqrt{x^2-2x+1} > 1.5$

14. $\sqrt[4]{x+3} < \sqrt{x+1}$

15. $\sqrt{x-4} > x-10$

16. $\sqrt{x+5} < \sqrt{x-3}$

17. $\sqrt{3-2x} > 4$

18. $\sqrt{4x-3} \leq 7$

Solve each radical equation or inequality by using a graph. Round solutions to the nearest tenth. Check your solution by any method.

19. $\sqrt{3x-4} < x-2$

20. $\sqrt{3-x} \geq x^2-1$

21. $\sqrt{x+4} > x^3$

22. $\sqrt{2x+3} < x-2$

23. $\sqrt[3]{x+7} < \sqrt{2x+3}$

24. $\sqrt{x^2+2} \geq \sqrt{7x-3}$

25. $\sqrt{x+9} \leq x^2-3x$

26. $\sqrt{x^3+1} > x+3$

27. $\sqrt{7x-1} \leq x-2$

Algebra 2

Practice

9.1 Introduction to Conic Sections

Solve each equation for *y*, graph the resulting equation, and identify the conic section.

1. $x^2 - 3y = 0$

2. $x^2 + y^2 = 400$

3. $9x^2 - y^2 = 9$

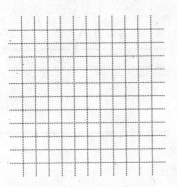

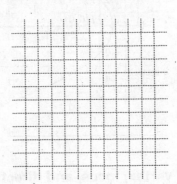

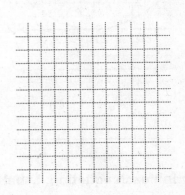

Find the distance between *P* and *Q*, and find the coordinates of *M*, the midpoint of $\overline{PQ}$. Give exact answers and approximate answers to the nearest hundredth when appropriate.

4. $P(0, 0)$ and $Q(5, 12)$

5. $P(4, 1)$ and $Q(12, -5)$

6. $P(12, 4)$ and $Q(-8, 2)$

7. $P(7.5, 3)$ and $Q(-1.5, 5)$

8. $P(-8, -8)$ and $Q(4, 4)$

9. $P(-1, -1)$ and $Q(1, 2)$

Find the center, circumference, and area of the circle whose diameter has the given endpoints.

10. $P(6, 20)$ and $Q(12, 8)$

11. $P(0, 0)$ and $Q(9, 40)$

12. $P(4, 16)$ and $Q(-4, 1)$

13. $P(3, 7)$ and $Q(4, -5)$

14. $P(10, 5)$ and $Q(20, 6)$

15. $P(-8, 8)$ and $Q(13, -3)$

NAME _____ CLASS _____ DATE _____

Practice
9.2 Parabolas

Write the standard equation for each parabola graphed below.

1.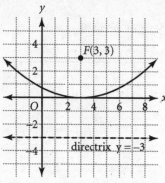

2.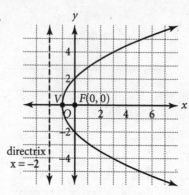

3.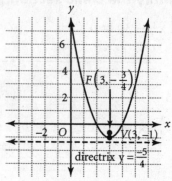

_____   _____   _____

Graph each equation. Label the vertex, focus, and directrix.

4. $x = \frac{1}{4}y^2$

5. $y - 4 = \frac{1}{4}(x - 1)^2$

6. $x - 1 = \frac{1}{8}(y - 2)^2$

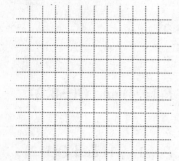   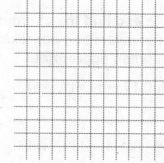   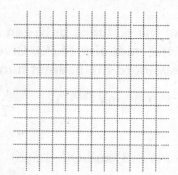

Write the standard equation for the parabola with the given characteristics.

7. vertex: $(0, 0)$; focus: $(0, 6)$ _____

8. vertex: $(10, 0)$; directrix: $x = 8$ _____

9. focus: $(3, 0)$; directrix: $x = -3$ _____

10. vertex: $(5, 2)$; directrix: $y = 1$ _____

11. vertex: $(6, -7)$; focus: $(4, -7)$ _____

12. focus: $(9, 5)$; directrix: $y = -5$ _____

56   Practice Workbook

Algebra 2

Copyright © by Holt, Rinehart and Winston. All rights reserved.

Practice
9.3 Circles

Write the standard equation for each circle graphed below.

1.

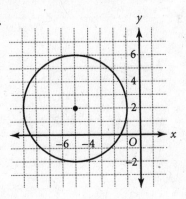

2.

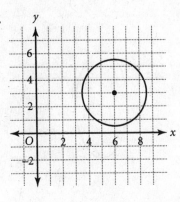

3.

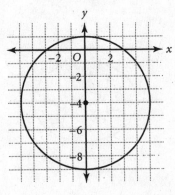

_____   _____   _____

Write the standard equation of a circle with the given radius and center.

4. $r = \dfrac{3}{4}$; $C(0, 0)$

5. $r = 2.5$; $C(-2, 1)$

6. $r = 24$; $C(-3, -3)$

_____   _____   _____

Graph each equation. Label the center and the radius.

7. $x^2 + y^2 = 256$

8. $x^2 + (y - 5)^2 = 16$

9. $(x - 3)^2 + (y + 3)^2 = 9$

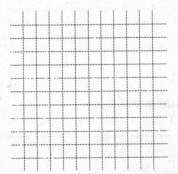   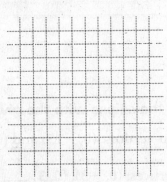   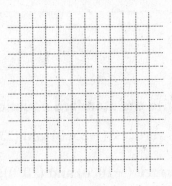

Write the standard equation for each circle. Then state the coordinates of its center, and give its radius.

10. $x^2 + y^2 - 10x - 16y + 88 = 0$

11. $x^2 + y^2 + 22x - 2y = -120$

_____   _____

Practice
9.4 *Ellipses*

Write the standard equation for each ellipse.

1.
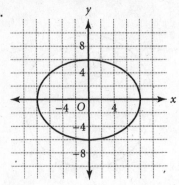

2.
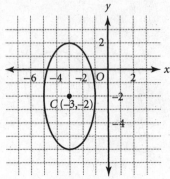

3.
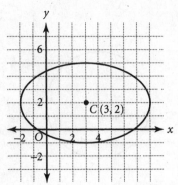

_____ _____ _____

Sketch the graph of each ellipse. Label the center, foci, vertices, and co-vertices.

4. $\dfrac{x^2}{4} + \dfrac{y^2}{81} = 1$

5. $\dfrac{x^2}{49} + \dfrac{(y-1)^2}{36} = 1$

6. $\dfrac{(x-4)^2}{9} + \dfrac{(y+3)^2}{25} = 1$

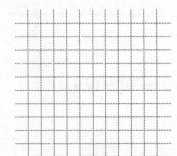
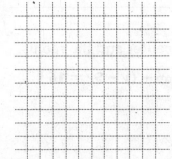
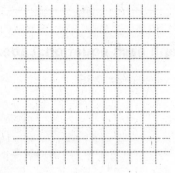

Write the standard equation for the ellipse with the given characteristics.

7. vertices: $(-25, 0)$ and $(25, 0)$; co-vertices: $(0, -15)$ and $(0, 15)$ _____

8. foci: $(-10, 0)$ and $(10, 0)$; co-vertices: $(0, -3)$, $(0, 3)$ _____

9. co-vertices: $(-20, 0)$ and $(20, 0)$; foci: $(0, -8)$ and $(0, 8)$ _____

10. An ellipse is defined by $x^2 + 4y^2 + 6x - 27 = 0$. Write the standard equation, and identify

the coordinates of the center, vertices, co-vertices, and foci. _____

_____

Practice
9.5 Hyperbolas

Write the standard equation for each hyperbola.

1.

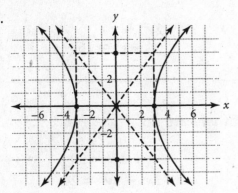

_____

2.

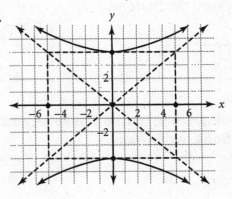

_____

Graph each hyperbola. Label the center, vertices, co-vertices, foci, and asymptotes.

3. $\dfrac{y^2}{9} - \dfrac{x^2}{25} = 1$

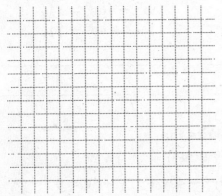

4. $\dfrac{(x-1)^2}{16} - \dfrac{(y-1)^2}{9} = 1$

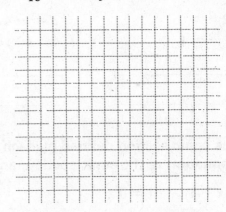

For Exercises 5–7, write the standard equation for the hyperbola with the given characteristics.

5. vertices: $\left(-\sqrt{10}, 0\right)$ and $\left(\sqrt{10}, 0\right)$; co-vertices: $\left(0, -\sqrt{15}\right)$ and $\left(0, \sqrt{15}\right)$ _____

6. foci: $(-5, -2)$ and $(5, -2)$; vertices: $(-3, 0)$ and $(3, 0)$ _____

7. center: $(1, 1)$; vertices: $(1, -4)$ and $(1, 6)$; co-vertices: $(13, 1)$ and $(-11, 1)$ _____

8. A hyperbola is defined by $x^2 - 4y^2 - 28x - 24y + 156 = 0$. Write the standard equation, and identify the coordinates of the center, vertices, co-vertices, and foci.

_____

_____

Practice

9.6 Solving Nonlinear Systems

Use the substitution method to solve each system. If there are no real solutions, write *none*.

1. $\begin{cases} y = x^2 + 5 \\ y = 5x + 1 \end{cases}$

2. $\begin{cases} y = x - 2 \\ y = x^2 \end{cases}$

3. $\begin{cases} y = x \\ x^2 + y^2 = 16 \end{cases}$

_____   _____   _____

Use the elimination method to solve each system. If there are no real solutions, write *none*.

4. $\begin{cases} x^2 + y^2 = 9 \\ 4x^2 - 9y^2 = 36 \end{cases}$

5. $\begin{cases} x^2 + y^2 = 8 \\ 2x^2 - 3y^2 = 1 \end{cases}$

6. $\begin{cases} x^2 + 2y^2 = 30 \\ 3x^2 - 5y^2 = 24 \end{cases}$

_____   _____   _____

_____   _____   _____

Solve each system by graphing. If there are no real solutions, write *none*.

7. $\begin{cases} x^2 + 2y^2 = 16 \\ 4x^2 + y^2 = 4 \end{cases}$

8. $\begin{cases} 25x^2 - 4y^2 = 100 \\ 4x^2 + 9y^2 = 36 \end{cases}$

9. $\begin{cases} 9x^2 - 16y^2 = 144 \\ 16x^2 + 9y^2 = 144 \end{cases}$

_____   _____   _____

_____   _____   _____

Classify the conic section defined by each equation. Write the standard equation of the conic section, and sketch the graph.

10. $x^2 - 14x - 4y + 61 = 0$

11. $4x^2 - 9y^2 - 40x + 72y - 80 = 0$

_____   _____

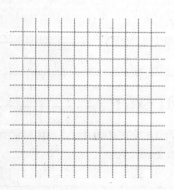   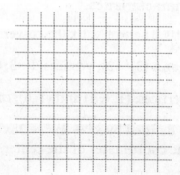

Practice
10.1 Introduction to Probability

Find the probability of each event.

1. A blue card is drawn at random from a bag containing 2 white cards, 1 red card, and 7 blue cards. _____

2. Frederique, who arrives home at 6:42 P.M., is home to receive a call that can come at any time between 6:40 and 6:50. _____

3. A letter chosen at random from the letters of the word *permutation* is a vowel. _____

4. A card chosen at random from a standard 52-card deck is a heart or a diamond. _____

5. A card chosen at random from a standard deck is not an 8 or an ace. _____

6. A number cube is rolled, and a number greater than 3 and less than 6 results. _____

7. A letter chosen at random from the alphabet is not one of the 5 standard vowels. _____

8. A point on a 12-inch ruler is chosen at random and is located within an inch of an end of the ruler. _____

A spinner is divided into three colored regions. You spin the spinner a total of 150 times. The results are recorded in the table. Find the experimental probability of each event.

green	42
yellow	65
pink	43

9. green _____

10. yellow _____ 11. pink _____

12. not pink _____ 13. not yellow _____

Find the number of possible license plate numbers (with no letters or digits excluded) for each of the following conditions:

14. 6 digits _____

15. 2 letters followed by 3 digits _____

16. 4 letters followed by 3 digits _____

17. 5 digits followed by 2 letters _____

18. 2 digits followed by 2 letters followed by 2 digits _____

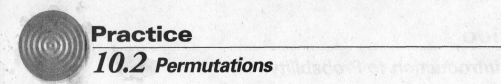

Practice
10.2 Permutations

Find the number of permutations of the first 7 letters of the alphabet for each situation.

1. taking all 7 letters at a time

_____

2. taking 5 letters at a time

_____

3. taking 4 letters at a time

_____

4. taking 3 letters at a time

_____

In how many ways can 12 books be displayed on a shelf if the given number of books are available?

5. 12 books

_____

6. 14 books

_____

7. 15 books

_____

8. 20 books

_____

Find the number of permutations of the letters in each word.

9. *geometry*

_____

10. *algebra*

_____

11. *addition*

_____

12. *calculus*

_____

13. *mathematics*

_____

14. *arithmetic*

_____

15. Lizette decorates windows for a department store. She plans to design a baby's room with a row of stuffed elephants and monkeys along one wall. If she has 8 identical elephants and 10 identical monkeys, in how many different ways can the stuffed animals be displayed? _____

16. The 6 candidates for a student government office are invited to speak at an election forum. In how many different orders can they speak? _____

17. Representatives from 8 schools are represented at a school newspaper workshop. In how many different ways can the 8 representatives be seated around a circular table? _____

18. Ten colleges are participating in a college fair. Booths will be positioned along one wall of a high school gymnasium. In how many different orders can the booths be arranged? _____

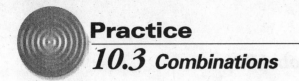

Practice
10.3 Combinations

Find the number of ways in which each committee can be selected.

1. a committee of 5 people from a group of 8 people _____

2. a committee of 2 people from a group of 16 people _____

3. a committee of 4 people from a group of 7 people _____

4. a committee of 8 people from a group of 15 people _____

5. a committee of 3 people from a group of 9 people _____

At a luncheon, guests are offered a selection of 4 different grilled vegetables and 5 different relishes. In how many ways can the following items be chosen?

6. 2 vegetables and 3 relishes

7. 3 vegetables and 2 relishes

_____   _____

8. 4 vegetables and 4 relishes

9. 3 vegetables and 3 relishes

_____   _____

A bag contains 8 white marbles and 7 blue marbles. Find the probability of selecting each combination.

10. 2 white and 3 blue   11. 3 white and 2 blue   12. 4 white and 1 blue

_____   _____   _____

Determine whether each situation involves a permutation or a combination.

13. A high school offers 5 foreign language programs. In how many ways can a student choose 2 programs?   _____

14. In how many ways can 20 members be chosen from a 60-member chorus to sing the national anthem at a graduation ceremony?   _____

15. In how many ways can a captain, co-captain, and team manager be chosen from among the 18 members of a volleyball team?   _____

16. First- through fourth-place prizes are to be awarded in an essay contest. In how many ways can the winners be selected from among 125 entries?   _____

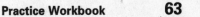

Practice

10.4 *Using Addition with Probability*

A card is drawn at random from a standard 52-card deck. Tell whether events *A* and *B* are inclusive or mutually exclusive. Then find *P(A or B)*.

1. *A*: The card is a heart.
   B: The card is an 8.

   _____

2. *A*: The card is a number less than 5.
   B: The card is a jack, a king, or a queen.

   _____

3. *A*: The card is black.
   B: The card is a number greater than 4.

   _____

4. *A*: The card is not a diamond.
   B: The card is a spade.

   _____

5. *A*: The card is red.
   B: The card is the ace of spades.

   _____

6. *A*: The card is a 2 or a 3.
   B: The card is not a heart.

   _____

A spinner is divided into 8 congruent regions numbered 1 through 8. The spinner is spun once. Find the probability of each event.

7. The number is even or divisible by 3.

   _____

8. The number is odd or greater than 7.

   _____

9. The number is less than 2 or greater than 6.

   _____

10. The number is odd or divisible by 4.

   _____

Two number cubes are rolled, and the numbers on the top faces are added. The table at right shows the possible outcomes. Find each probability.

+	1	2	3	4	5	6
1	2	3	4	5	6	7
2	3	4	5	6	7	8
3	4	5	6	7	8	9
4	5	6	7	8	9	10
5	6	7	8	9	10	11
6	7	8	9	10	11	12

11. The sum is odd *or* greater than 11.

   _____

12. The sum is less than 6 *or* greater than 10.

   _____

13. The sum is even *or* less than 5.

   _____

14. The sum is less than 8 *or* a multiple of 6.

   _____

15. The sum is less than 4 *or* a multiple of 5.

   _____

Practice

10.5 Independent Events

Events *D*, *E*, *F* and *G* are independent, and *P*(*D*) = 0.2, *P*(*E*) = 0.1, *P*(*F*) = 0.4, and *P*(*G*) = 0.25. Find the probability of each combination of events.

1. *P*(*D and E*) _____

2. *P*(*D and F*) _____

3. *P*(*E and F*) _____

4. *P*(*D and G*) _____

5. *P*(*D and E and F*) _____

6. *P*(*E and F and G*) _____

A bag contains 3 white marbles, 2 red marbles, and 7 blue marbles. A marble is picked at random and is replaced. Then a second marble is picked at random. Find each probability.

7. Both marbles are blue. _____

8. The first marble is white and the second marble is red. _____

9. The first marble is white and the second marble is not white. _____

10. Neither marble is red. _____

11. The first marble is blue and the second marble is red. _____

A number cube is rolled twice. On each roll, the number on the top face is recorded. Find the probability of each event.

12. The first number is greater than 5 and the second is less than 3. _____

13. Both numbers are greater than 4. _____

14. The first number is even and the second number is odd. _____

15. Both numbers are less than 2. _____

16. Neither number is greater than 4. _____

A number cube is rolled, and two coins are tossed. Find the probability of each event.

17. The number on the cube is 2 and both coins are heads. _____

18. The number on the cube is even, one coin shows heads, and one shows tails. _____

19. The number on the cube is greater than 4 and both coins are tails. _____

20. The number on the cube is greater than 2 and the coins show different sides. _____

Practice
10.6 Dependent Events and Conditional Probability

Two number cubes are rolled, and the first cube shows 6. Find the probability of each event below.

1. The sum is 9. _____

2. Both numbers are even. _____

3. The sum is greater than 8. _____

4. The sum is greater than 9 and less than 12. _____

A spinner that is divided into 8 congruent regions, numbered 1 through 8, is spun once. Let *A* be the event "even" and let *B* be the event "6." Find each of the following probabilities.

5. $P(A)$

6. $P(B)$

7. $P(A \text{ and } B)$

_____   _____   _____

8. $P(A \text{ or } B)$

9. $P(A|B)$

10. $P(B|A)$

_____   _____   _____

A spinner that is divided into 5 congruent regions, numbered 1 through 5, is spun once. Let *A* be the event "odd" and let *B* be the event "less than 3." Find each of the following probabilities.

11. $P(A)$

12. $P(B)$

13. $P(A \text{ and } B)$

_____   _____   _____

14. $P(A \text{ or } B)$

15. $P(A|B)$

16. $P(B|A)$

_____   _____   _____

Let *A* and *B* represent events.

17. Given $P(A \text{ and } B) = 0.25$ and $P(A) = 0.4$, find $P(B|A)$. _____

18. Given $P(A \text{ and } B) = \frac{3}{5}$ and $P(A) = \frac{2}{3}$, find $P(B|A)$. _____

19. Given $P(B|A) = \frac{4}{5}$ and $P(A) = \frac{5}{8}$, find $P(A \text{ and } B)$. _____

20. Given $P(B|A) = 0.4$ and $P(A) = 0.16$, find $P(A \text{ and } B)$. _____

21. Given $P(B|A) = 0.5$ and $P(A \text{ and } B) = 0.2$, find $P(A)$. _____

22. Given $P(B|A) = 0.8$ and $P(A \text{ and } B) = 0.45$, find $P(A)$. _____

Algebra 2

Practice

10.7 *Experimental Probability and Simulation*

Use a simulation with 10 trials to find an estimate for each probability.

1. In 4 tosses of a coin, heads appears exactly 3 times.

Trial	Result
1	
2	
3	
4	
5	
6	
7	
8	
9	
10	

estimate: _____

2. In 5 tosses of a coin, tails appears more than 2 times.

Trial	Result
1	
2	
3	
4	
5	
6	
7	
8	
9	
10	

estimate: _____

3. In 4 rolls of a number cube, 3 appears twice.

Trial	Result
1	
2	
3	
4	
5	
6	
7	
8	
9	
10	

estimate: _____

Of 100 motorists observed at an intersection, 26 turned left, 47 went straight, and 27 turned right. Use a simulation with 10 trials to find an estimate for each probability.

4. Exactly 2 of every 4 motorists turn right.

Trial	Result
1	
2	
3	
4	
5	
6	
7	
8	
9	
10	

estimate: _____

5. At least 3 of every 4 motorists go straight.

Trial	Result
1	
2	
3	
4	
5	
6	
7	
8	
9	
10	

estimate: _____

6. Fewer than 2 of every 4 motorists turn left.

Trial	Result
1	
2	
3	
4	
5	
6	
7	
8	
9	
10	

estimate: _____

Practice

11.1 Sequences and Series

Write the first six terms of each sequence.

1. $b_n = 2.5n$

2. $f_n = \frac{1}{2}n - \frac{1}{2}$

_____

3. $t_n = n^2 + 12$

4. $a_1 = 20; a_n = 3a_{n-1} + 10$

_____

5. $a_1 = 1; a_n = a_{n-1} + 100$

6. $a_1 = -5; a_n = 3a_{n-1}$

_____

For each sequence below, write a recursive formula, and find the next three terms.

7. $1, 11, 121, 1331, \ldots$

8. $81, 78, 75, 72, \ldots$

9. $2, -6, 18, -54, \ldots$

_____

10. $\frac{1}{4}, -1, 4, -16, \ldots$

11. $2, 11, 38, 119, \ldots$

12. $-2, -14, -74, -374, \ldots$

_____

Write the terms of each series. Then evaluate.

13. $\displaystyle\sum_{n=1}^{7} 4.5n$

14. $\displaystyle\sum_{j=1}^{5} (j^2 + 8j + 2)$

_____

Evaluate.

15. $\displaystyle\sum_{m=1}^{6} 10m$

16. $\displaystyle\sum_{n=1}^{8} (2n - 12)$

17. $\displaystyle\sum_{j=1}^{5} (j - 3)^2$

_____

18. $\displaystyle\sum_{k=1}^{10} (2k - 0.5)^2$

19. $\displaystyle\sum_{a=1}^{6} (4a^2 + 3a - 5)$

20. $\displaystyle\sum_{n=1}^{12} (3.5n^2 - 5n + 2.2)$

_____

Practice
11.2 Arithmetic Sequences

Based on the terms given, state whether or not each sequence is arithmetic. If so, identify the common difference, d.

1. $15, 18, 21, 24, \ldots$

2. $2, 5, 10, 17, \ldots$

3. $7.2, 9.7, 12.2, 14.7, \ldots$

_____

_____

_____

4. $4, 6, 9, 13.5, \ldots$

5. $1, 1\frac{3}{5}, 2\frac{1}{5}, 2\frac{4}{5}, \ldots$

6. $8, 5.7, 3.4, 1.1, \ldots$

_____

_____

_____

Write an explicit formula for the nth term of each arithmetic sequence.

7. $16, 7, -2, -11, \ldots$

8. $-15, -7, 1, 9, \ldots$

9. $13, 16, 19, 22, \ldots$

_____

_____

_____

10. $-25, -13, -1, 11, \ldots$

11. $9, 20, 31, 42, \ldots$

12. $8.6, 7.3, 6, 4.7, \ldots$

_____

_____

_____

List the first four terms of each arithmetic sequence.

13. $t_1 = 50; t_n = t_{n-1} + 100$

14. $t_1 = 7.5; t_n = t_{n-1} + 2.5$

15. $t_1 = -20; t_n = t_{n-1} + 8$

_____

_____

_____

16. $t_n = 40n - 15$

17. $t_n = 0.5n + 8$

18. $t_n = -12n - 3$

_____

_____

_____

Find the indicated arithmetic means.

19. 3 arithmetic means between -12 and 16

20. 4 arithmetic means between 40 and 100

_____

_____

21. 2 arithmetic means between 50 and 86

22. 3 arithmetic means between 7 and 21

_____

_____

23. 3 arithmetic means between 40 and 16

24. 4 arithmetic means between -8 and 22

_____

_____

Practice
11.3 Arithmetic Series

Use the formula for an arithmetic series to find each sum.

1. $62 + 66 + 70 + 74 + 78$ _____

2. $\frac{1}{2} + 2\frac{1}{2} + 3\frac{1}{2} + 4\frac{1}{2} + 5\frac{1}{2}$ _____

3. $-30 - 27 - 24 - 21 - \cdots - 3$ _____

4. $110 + 125 + 140 + \cdots + 305$ _____

5. $14 + 17 + 20 + \cdots + 65$ _____

6. $33 + 38 + 43 + \cdots + 123$ _____

Find each sum.

7. the sum of the first 225 natural numbers _____

8. the sum of the first 15 multiples of 3 _____

9. the sum of the first 25 multiples of 4 _____

10. the sum of the multiples of 5 from 75 to 315, inclusive _____

11. the sum of the multiples of 7 from 84 to 371, inclusive _____

For each arithmetic series, find S_{22}.

12. $-6 + (-4) + (-2) + 0 + \cdots$

13. $3 + 7 + 11 + 15 + \cdots$

_____

_____

14. $-24 + (-21) + (-18) + (-15) + \cdots$

15. $3 + 3\frac{3}{4} + 4\frac{1}{2} + 5\frac{1}{4} + \cdots$

_____

_____

16. $18 + 8 + (-2) + (-12) + \cdots$

17. $3\sqrt{5} + 5\sqrt{5} + 7\sqrt{5} + 9\sqrt{5} + \cdots$

_____

_____

Evaluate.

18. $\sum_{n=1}^{6} (2n + 7)$

19. $\sum_{j=1}^{8} (-3j - 3)$

20. $\sum_{k=1}^{10} (10k + 4)$

_____

_____

_____

21. $\sum_{m=1}^{12} (-7 + 4m)$

22. $\sum_{b=1}^{15} (13 + 5b)$

23. $\sum_{i=1}^{9} (-8i + 1)$

_____

_____

_____

Practice

11.4 Geometric Sequences

Determine whether each sequence is geometric. If so, identify the common ratio, *r*, and give the next three terms.

1. $9, 25, 49, 81, \ldots$

2. $200, 80, 32, 12.8, \ldots$

3. $66\frac{2}{3}, -40, 24, -14\frac{2}{5}, \ldots$

4. $12, 18, 27, 40.5, \ldots$

5. $54, 36, 24, 16, \ldots$

6. $1, \sqrt{2}, \sqrt{3}, 2, \ldots$

List the indicated terms of each geometric sequence.

7. $t_1 = 18; t_n = -2t_{n-1};$ first 4 terms

8. $t_1 = -4; t_n = 2.5t_{n-1};$ first 5 terms

9. $t_1 = 20; t_n = 0.5t_{n-1};$ first 4 terms

10. $t_2 = 40; t_4 = 2.5; t_5$

11. $t_2 = 16; t_6 = 64; t_5$

12. $t_3 = 20\frac{1}{4}; t_5 = 182\frac{1}{4}; t_7$

Write an explicit formula for the *n*th term of each geometric sequence.

13. $250, 100, 40, 16, \ldots$

14. $-30, 6, -1.2, 0.24, \ldots$

15. $40, 32, 25.6, 20.48, \ldots$

16. $2, 5, 12.5, 31.25, \ldots$

17. $20, 5, 1\frac{1}{4}, \frac{5}{16}, \ldots$

18. $1.5, -9, 54, -324, \ldots$

19. Find 2 geometric means between 7 and 875.

20. Find 2 geometric means between -28 and -3.5.

21. Find 3 geometric means between 12 and 3072.

22. Find 3 geometric means between 12.5 and 25.92.

23. Find 3 geometric means between 12 and 7500.

24. Find 4 geometric means between 4 and 972.

Practice

11.5 Geometric Series and Mathematical Induction

Find each sum. Round answers to the nearest tenth, if necessary.

1. S_{20} for the geometric series $4 + 12 + 36 + 108 + \cdots$ _____

2. S_{15} for the geometric series $72 + 12 + 2 + \frac{1}{3} + \cdots$ _____

3. S_6 for the series related to the geometric sequence $7, -14, 28, -56, \ldots$ _____

4. $\frac{7}{5} + \frac{7}{25} + \frac{7}{125} + \frac{7}{625} + \frac{7}{3125}$ _____

5. $1.3 - 5.2 + 20.8 - 83.2 + 332.8 - 1331.2$ _____

For Exercises 6–9, refer to the geometric sequence 3, 6, 12, 24, . . .

6. Find t_{12}.   7. Find t_{20}.   8. Find S_{12}.   9. Find S_{20}.

_____   _____   _____   _____

Evaluate. Round answers to the nearest hundredth, if necessary.

10. $\displaystyle\sum_{k=1}^{6} 6(2^{k-1})$ 11. $\displaystyle\sum_{n=1}^{10} 4.8^{n-1}$ 12. $\displaystyle\sum_{j=1}^{12} 5(0.25^k)$ 13. $\displaystyle\sum_{m=1}^{15} \frac{2}{3}(3^m)$

_____   _____   _____   _____

14. $\displaystyle\sum_{p=1}^{6} 2(3)^{p-1}$ 15. $\displaystyle\sum_{t=1}^{10} 3(-1)^{t+2}$ 16. $\displaystyle\sum_{k=1}^{10} 5^k$ 17. $\displaystyle\sum_{k=1}^{12} 3(2)^k$

_____   _____   _____   _____

Use mathematical induction to prove that the statement is true for every natural number, _n_.

18. $1^3 + 2^3 + 3^3 + \cdots + n^3 = \dfrac{n^2(n+1)^2}{4}$ _____

_____

_____

_____

_____

_____

_____

Practice
11.6 Infinite Geometric Series

Find the sum of each infinite geometric series, if it exists.

1. $60 + 84 + 117.6 + 164.64 + \cdots$

2. $\frac{4}{5} + \frac{4}{15} + \frac{4}{45} + \frac{4}{135} + \cdots$

3. $\frac{7}{8} + \frac{7}{12} + \frac{7}{18} + \frac{7}{27} + \cdots$

4. $5 + 4 + 3.2 + 2.56 + \cdots$

Find the sum of each infinite geometric series, if it exists.

5. $\sum_{n=1}^{\infty} 0.8^n$

6. $\sum_{m=1}^{\infty} \left(\frac{11}{9}\right)^{m-1}$

7. $\sum_{k=1}^{\infty} 11 \cdot \left(\frac{1}{9}\right)^{k-1}$

8. $\sum_{j=1}^{\infty} 0.75^j$

9. $\sum_{t=1}^{\infty} 0.45^{t-1}$

10. $\sum_{x=1}^{\infty} 0.92^x$

11. $\sum_{k=1}^{\infty} 7.3^{k-1}$

12. $\sum_{b=1}^{\infty} 49(0.02)^{b-1}$

13. $\sum_{n=1}^{\infty} 20(0.1)^{n-1}$

Write each decimal as a fraction in simplest form.

14. $0.\overline{1}$ _____

15. $0.\overline{37}$ _____

16. $0.\overline{49}$ _____

17. $0.\overline{753}$ _____

18. $0.\overline{225}$ _____

19. $0.\overline{370}$ _____

Write an infinite geometric series that converges to the given number.

20. $0.0707070707\ldots$ _____

21. $0.9393939393\ldots$ _____

22. $0.1515151515\ldots$ _____

23. $0.358358358\ldots$ _____

24. $0.011011011\ldots$ _____

25. $0.445445445\ldots$ _____

Practice
11.7 Pascal's Triangle

State the location of each entry in Pascal's triangle. Then give the value of each expression.

1. $_7C_5$

2. $_6C_3$

3. $_8C_6$

4. $_{10}C_5$

5. $_{13}C_{10}$

6. $_{12}C_5$

Find the indicated entries in Pascal's triangle.

7. fourth entry,
   row 10

8. seventh entry,
   row 13

9. ninth entry,
   row 15

10. third entry,
    row 18

A fair coin is tossed the indicated number of times. Find the probability of each event.

11. 5 tosses; exactly 3 heads _____

12. 6 tosses; no more than 3 heads _____

13. 10 tosses; exactly 1 head _____

14. 8 tosses; fewer than 5 heads _____

15. 5 tosses; no fewer than 3 heads _____

16. 7 tosses; 2 *or* 3 heads _____

A student guesses the answers to 6 questions on a true-false quiz. Find the probability that the indicated number of guesses are correct.

17. exactly 4

18. fewer than 5

19. no more than 2

20. exactly 5

21. at least 3

22. at least 4

Practice
11.8 The Binomial Theorem

Expand each binomial raised to a power.

1. $(s + t)^5$ _____

2. $(b + w)^6$ _____

For Exercises 3–5, refer to the expansion of $(r + s)^{15}$.

3. How many terms are in the expansion? _____

4. What is the exponent of r in the term that contains s^{12}? _____

5. Write the term that contains r^5. _____

Expand each binomial.

6. $(5x + y)^5$

_____

7. $\left(\frac{1}{2}w + 2z\right)^4$

_____

8. $\left(\frac{3}{4}a - d\right)^5$

_____

9. $(2m - 3q)^6$

_____

Ty Cobb was the American League batting champion for 12 of his 24 years in baseball. His lifetime batting average was 0.367. Use the Binomial Theorem to find each theoretical probability for Ty Cobb.

10. exactly 3 hits in 5 at bats _____

11. at least 3 hits in 5 at bats _____

12. no more than 2 hits in 5 at bats _____

13. exactly 4 hits in 6 at bats _____

14. fewer than 3 hits in 6 at bats _____

Practice
12.1 *Measures of Central Tendency*

Find the mean, median, and mode of each data set. Round answers to the nearest thousandth, when necessary.

1. 62, 54, 63, 92, 62, 79, 54, 62 _____

2. 12, 28, 40, 22, 33, 28, 9 _____

3. 3.6, 6.3, 1.3, 3.6, 1.0, 5.9 _____

4. 277, 725, 920, 835, 255, 725 _____

5. 1828, 1008, 1600, 7309, 2215, 1600 _____

Find the mean, median, and mode of the data, and compare them.

6. percent of total social welfare expenditures for education
in 1985–1992: 22.8, 23.2, 24.5, 24.8, 25.0, 25.0, 24.0, 23.1

_____

_____

Make a frequency table for the data, and find the mean.

7. ages (in years) of members of the swim team:
14, 15, 17, 17, 18, 16, 15, 14, 16, 17, 17, 18, 17,
16, 16, 15, 14, 17

mean: _____

Age (years)	Tally	Frequency
14		
15		
16		
17		
18		

Make a grouped frequency table for the data, and estimate the mean.

8. number of books read by the
members of a class in the past year:
5, 4, 12, 22, 30, 5, 7, 3, 1, 10, 12, 16,
26, 15, 10, 2, 5, 3, 10, 21

estimated mean: _____

Number of books	Class mean	Frequency	Product
1–5			
6–10			
11–15			
16–20			
21–25			
26–30			

Practice

12.2 *Stem-and-Leaf Plots, Histograms, and Circle Graphs*

**Make a stem-and-leaf plot for each data set. Then find the median
and the mode, and describe the distribution of the data.**

1. 40, 64, 54, 38, 42, 45, 33, 37, 56, 58, 64

Stem	Leaf

2. 3.6, 4.8, 3.9, 1.7, 4.3, 2.3, 4.8, 3.1, 4.0, 2.3

Stem	Leaf

_____

Make a frequency table and a histogram for the data.

3. 1.0, 1.3, 1.1, 1.4, 1.4, 1.2, 1.1, 1.0, 1.0, 1.3, 1.3, 1.4, 1.3, 1.2, 1.0, 1.3, 1.4, 1.2, 1.0, 1.3

Number	Frequency

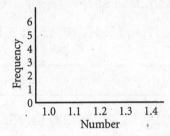

Complete the table, and make a relative frequency histogram for the data.

4.

Number	Frequency	Relative frequency
10	8	
11	12	
12	7	
13	11	
14	12	

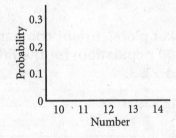

Make a circle graph for the data.

5.

Motor Vehicle Registration by Type, 1994			
Passenger cars	Motorcycles	Buses	Trucks
33.2%	0.9%	15.9%	50%

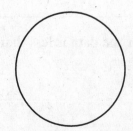

Practice
12.3 Box-and-Whisker Plots

Find the quartiles, the range, and the interquartile range for each data set.

1. 9, 5, 2, 8, 2, 8, 7, 8, 3, 2, 8, 1, 9, 1, 5, 8, 9, 7, 9, 6

_____

2. 13, 14, 15, 19, 16, 19, 8, 17, 10, 7, 5, 18, 10, 16, 17, 12

_____

3. 35.1, 40.3, 13.8, 15.3, 42.7, 40.8, 15.5, 38.5, 28.4, 11.0, 11.7, 12.1, 23.9, 8.9, 24.0

_____

Find the minimum and maximum values, quartiles, range, and interquartile range for each data set. Then make a box-and-whisker plot for each data set.

4. 17, 14, 5, 8, 15, 4, 11, 6, 13, 17, 17, 13, 7, 9, 5, 3

_____

_____

5. 35, 35, 23, 20, 29, 13, 26, 21, 39, 22, 14, 35, 10, 16, 36

_____

_____

The box-and-whisker plots at right compare birth rates (per 1000 population) for the fifty states for 1992 and 1993.

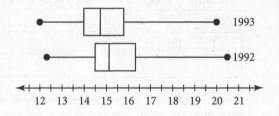

6. Which set of data has

   the greater median? _____

7. Does it appear that, in general, the birth rates increased or decreased from 1992 to 1993?

_____

8. What percent of the data is less than Q_3 for each year? _____

Practice

12.4 *Measures of Dispersion*

Find the range and mean deviation for each data set.

1. 24, 20, 38, 36, 52

_____

2. 12, 11, 15, 18, 22, 30

_____

3. 71, 56, 88, 82, 40, 95

_____

4. 120, 142, 167, 188, 167, 200

_____

5. 5.8, 3.4, 7.2, 10.5, 8.6

_____

6. 38, 52, 40, 61, 53, 90, 100

_____

Find the variance and standard deviation for each data set.

7. 13, 13, 17, 11, 22, 20

_____

8. 82, 44, 67, 52, 120

_____

9. 1215, 1805, 1715, 2010, 1875

_____

10. 12, 14.5, 18, 16, 11.5, 15

_____

11. 30, 40.2, 40.8, 22.6, 18

_____

12. 19.4, 19, 19.2, 19.6, 19.8, 19

_____

The table shows the winning scores in the United States Women's Open Golf Championships from 1977 to 1996. Refer to the data in the table for Exercises 13–16.

292	289	284	280	279	283	290	290	280	287
285	277	278	284	283	280	280	277	278	272

13. Find the range.

_____

14. Find the mean deviation.

_____

15. Find the variance.

_____

16. Find the standard deviation.

_____

Practice

12.5 Binomial Distributions

A coin is flipped 6 times. Find the probability of each event.

1. exactly 1 head

_____

2. exactly 5 heads

_____

3. more than 3 heads

_____

4. fewer than 2 heads

_____

A spinner is divided into 5 congruent segments, each labeled with one of the letters *A–E*. Find the probability of each event.

5. exactly 3 *A*s in 3 spins

_____

6. fewer than 2 *B*s in 4 spins

_____

7. exactly 4 *C*s in 5 spins

_____

8. fewer than 3 *D*s in 5 spins

_____

9. more than 3 *E*s in 5 spins

_____

10. exactly 3 *A*s in 10 spins

_____

At one university, the probability that an entering student will graduate is 40%. Find the probability of each event.

11. Exactly 4 out of 5 randomly selected entering students will graduate. _____

12. Fewer than 3 out of 5 randomly selected entering students will graduate. _____

13. Exactly 2 out of 6 randomly selected entering students will graduate. _____

14. More than 3 out of 6 randomly selected entering students will graduate. _____

The probability that any given person is left-handed is about 10%. Find each of the following probabilities:

15. Exactly 3 out of 7 randomly selected people are left-handed. _____

16. More than 3 out of 7 randomly selected people are left-handed. _____

17. Fewer than 4 out of 7 randomly selected people are left-handed. _____

18. Exactly 4 out of 10 randomly selected people are left-handed. _____

Practice
12.6 Normal Distributions

Let x be a random variable with a standard normal distribution.
Use the area table for a standard normal curve, given on page 807
of the textbook, to find each probability.

1. $P(x \geq 0)$

2. $P(x \leq 1.2)$

3. $P(x \geq -1.8)$

_____

4. $P(0 \leq x < 0.4)$

5. $P(0 \leq x \leq 2.0)$

6. $P(-0.2 \leq x \leq 0)$

_____

7. $P(1.0 \leq x \leq 2.0)$

8. $P(-0.2 \leq x \leq 0.2)$

9. $P(-0.4 \leq x \leq 1.2)$

_____

**The time required to finish a given test is normally distibuted
with a mean of 40 minutes and a standard deviation of 8 minutes.**

10. What is the probability that a student chosen at randon will finish
in less than 32 minutes? _____

11. What is the probability that a student chosen at random will take
more than than 56 minutes to finish? _____

12. What is the probability that a student chosen at random will take
between 24 minutes and 48 minutes? _____

**The owners of a restaurant determine that the number of
minutes that a customer waits to be served is normally
distributed with a mean of 6 minutes and a standard deviation
of 2 minutes.**

13. What is the probability that a randomly selected customer will be
served in less than 4 minutes? _____

14. During a survey, 500 customers are served. How many would you
expect to be served in less than 8 minutes? _____

15. If 1000 customers are served, how many would you expect to
wait between 4 minutes and 10 minutes? _____

NAME _____ CLASS _____ DATE _____

Practice
13.1 Right-Triangle Trigonometry

Refer to the triangle at right to find each value listed.
Give exact answers and answers rounded to the
nearest ten-thousandth.

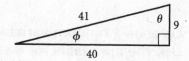

1. sin θ

_____

2. cos θ

_____

3. tan θ

_____

4. sin φ

_____

5. cos φ

_____

6. tan φ

_____

7. sec θ

_____

8. csc φ

_____

9. cot φ

_____

Solve each triangle. Round angle measures to the nearest degree
and side lengths to the nearest tenth.

10.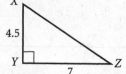

$XZ \approx$ _____

$m\angle X \approx$ _____

$m\angle Z \approx$ _____

11.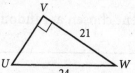

$UV \approx$ _____

$m\angle U \approx$ _____

$m\angle W \approx$ _____

12.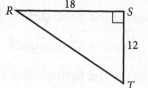

$RT \approx$ _____

$m\angle R \approx$ _____

$m\angle T \approx$ _____

13.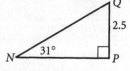

$m\angle Q \approx$ _____

$QN \approx$ _____

$NP \approx$ _____

14.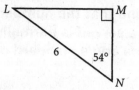

$m\angle L \approx$ _____

$LM \approx$ _____

$MN \approx$ _____

15.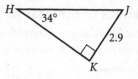

$m\angle J \approx$ _____

$HJ \approx$ _____

$HK \approx$ _____

Algebra 2

Practice
13.2 Angles of Rotation

For each angle below, find all coterminal angles, θ, such that −360° < θ < 360°. Then find the corresponding reference angle.

1. 47° _____

2. −123° _____

3. 218° _____

4. 512° _____

5. −222° _____

6. 307° _____

7. 1122° _____

8. −185° _____

9. 645° _____

Find the reference angle.

10. 105° _____

11. −213° _____

12. 715° _____

13. −144° _____

14. 860° _____

15. −72° _____

16. −2° _____

17. 1000° _____

18. −420° _____

Find the exact values of the six trigonometric functions of θ, given each point on the terminal side of θ in standard position.

19. (12, 8)

$\sin \theta =$ _____

$\cos \theta =$ _____

$\tan \theta =$ _____

$\csc \theta =$ _____

$\sec \theta =$ _____

$\cot \theta =$ _____

20. (−5, 10)

$\sin \theta =$ _____

$\cos \theta =$ _____

$\tan \theta =$ _____

$\csc \theta =$ _____

$\sec \theta =$ _____

$\cot \theta =$ _____

21. (4, 9)

$\sin \theta =$ _____

$\cos \theta =$ _____

$\tan \theta =$ _____

$\csc \theta =$ _____

$\sec \theta =$ _____

$\cot \theta =$ _____

Given the quadrant of θ in standard position and a trigonometric function value of θ, find exact values for the indicated trigonometric function.

22. IV, $\sin \theta = -\frac{3}{5}$; $\tan \theta$

23. I, $\tan \theta = \frac{5}{8}$; $\csc \theta$

24. II, $\cos \theta = -\frac{5}{8}$; $\sin \theta$

25. III, $\csc \theta = -1.25$; $\tan \theta$

26. II, $\cot \theta = -2.4$; $\sin \theta$

27. IV, $\sec \theta = \frac{4}{3}$; $\cot \theta$

Practice

13.3 Trigonometric Functions of Any Angle

Point *P* is located at the intersection of a circle with a radius of *r* and the terminal side of angle θ. Find the exact coordinates of *P*.

1. $\theta = 45°; r = 5$

2. $\theta = 60°; r = 12$

3. $\theta = -120°; r = 15$

_____ _____ _____

4. $\theta = 330°; r = 40$

5. $\theta = 135°; r = 10$

6. $\theta = 750°; r = 4$

_____ _____ _____

Point *P* is located at the intersection of the unit circle and the terminal side of angle θ in standard position. Find the coordinates of *P* to the nearest thousandth.

7. $\theta = 42°$

8. $\theta = 129°$

9. $\theta = 244°$

_____ _____ _____

10. $\theta = 305°$

11. $\theta = -41°$

12. $\theta = -105°$

_____ _____ _____

Find the exact values of the sine, cosine, and tangent of each angle.

13. $2160°$ 14. $315°$ 15. $-240°$ 16. $1770°$

sin: _____   sin: _____   sin: _____   sin: _____

cos: _____   cos: _____   cos: _____   cos: _____

tan: _____   tan: _____   tan: _____   tan: _____

Find each trigonometric function value. Give exact answers.

17. $\sin 420°$ 18. $\cos(-150°)$ 19. $\csc(-480°)$

_____ _____ _____

20. $\tan 300°$ 21. $\cos 1035°$ 22. $\sin 1470°$

_____ _____ _____

23. $\cot(-120°)$ 24. $\tan 495°$ 25. $\csc 210°$

_____ _____ _____

Practice
13.4 Radian Measure and Arc Length

Convert each degree measure to radian measure. Give exact answers.

1. $135°$ 2. $300°$ 3. $36°$ 4. $150°$

_____   _____   _____   _____

5. $105°$ 6. $-85°$ 7. $70°$ 8. $75°$

_____   _____   _____   _____

Convert each radian measure to degree measure. Give answers to the nearest hundredth of a degree.

9. $\frac{5\pi}{2}$ radians   10. $\frac{11\pi}{12}$ radians   11. $\frac{7\pi}{9}$ radians   12. $\frac{13\pi}{12}$ radians

_____   _____   _____   _____

13. 8.25 radians   14. 1.8 radians   15. 3 radians   16. 0.5 radian

_____   _____   _____   _____

A circle has a diameter of 20 feet. For each central angle measure below, find the length in feet of the arc intercepted by the angle.

17. $\frac{3\pi}{4}$ radians   18. $\frac{\pi}{12}$ radian   19. $\frac{2\pi}{3}$ radians   20. $\frac{\pi}{6}$ radian

_____   _____   _____   _____

21. 2.5 radians   22. 4 radians   23. 7.3 radians   24. 10 radians

_____   _____   _____   _____

Evaluate each expression. Give exact values.

25. $\sin 3\pi$ 26. $\cos \frac{2\pi}{3}$ 27. $\tan \frac{5\pi}{3}$ 28. $\csc\left(-\frac{\pi}{2}\right)$

_____   _____   _____   _____

29. $\tan \frac{13\pi}{6}$ 30. $\sin \frac{7\pi}{4}$ 31. $\cos \frac{5\pi}{2}$ 32. $\sec \frac{5\pi}{3}$

_____   _____   _____   _____

Practice

13.5 Graphing Trigonometric Functions

Complete the table of values for the function and graph the function along with its parent function.

1.

θ	0°	30°	45°	60°	90°
cos 2θ					

120°	135°	150°	180°

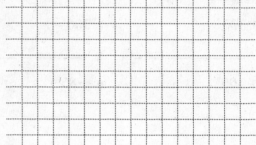

Identify the amplitude, if it exists, and the period of each function.

2. $y = 4.5 \cos 2\theta$

3. $y = 3 \tan\left(x - \frac{\pi}{2}\right) + 1$

4. $y = 1.2 \cos(x + \pi)$

_____   _____   _____

Identify the phase shift and vertical translation of each function from its parent function. Then graph at least one period of the function for 0° ≤ θ ≤ 360°, or 0 ≤ x ≤ 2π.

5. $y = 2 \cos(\theta - 45°) + 1.5$

6. $y = \sin 2(x + \pi) - 1$

_____   _____

_____   _____

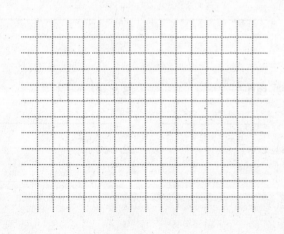 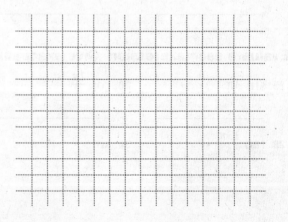

Practice
13.6 *Inverses of Trigonometric Functions*

Evaluate each trigonometric expression. Give exact answers in degrees.

1. $\mathrm{Sin}^{-1}\left(-\frac{\sqrt{3}}{2}\right)$

_____

2. $\mathrm{Cos}^{-1}\left(-\frac{1}{2}\right)$

_____

3. $\mathrm{Tan}^{-1}\,0$

_____

4. $\mathrm{Cos}^{-1}\,1$

_____

5. $\mathrm{Sin}^{-1}\left(-\frac{\sqrt{2}}{2}\right)$

_____

6. $\mathrm{Tan}^{-1}(-\sqrt{3})$

_____

Evaluate each trigonometric expression.

7. $\cos(\mathrm{Sin}^{-1}\,1)$

_____

8. $\tan\left(\mathrm{Cos}^{-1}\left(-\frac{\sqrt{2}}{2}\right)\right)$

_____

9. $\sin(\mathrm{Tan}^{-1}\sqrt{3})$

_____

10. $\mathrm{Cos}^{-1}\left(\sin\frac{\pi}{3}\right)$

_____

11. $\mathrm{Tan}^{-1}(\cos(-720°))$

_____

12. $\mathrm{Sin}^{-1}\left(\tan\frac{5\pi}{4}\right)$

_____

Find each value. Give answers in radians, rounded to the nearest ten-thousandth.

13. $\mathrm{Tan}^{-1}\,40.2356$

_____

14. $\mathrm{Sin}^{-1}\,0.0345$

_____

15. $\mathrm{Cos}^{-1}\,(-0.8114)$

_____

16. $\mathrm{Cos}^{-1}\,0.7756$

_____

17. $\mathrm{Tan}^{-1}\,(-38.2004)$

_____

18. $\mathrm{Sin}^{-1}\,(-0.5454)$

_____

Use inverse trigonometric functions to solve each problem.

19. A ramp that is 18 feet long rises to a loading platform that is 3 feet above the ground. Find, to the nearest tenth of a degree, the angle that the ramp makes with the ground.   _____

20. At one point in the day, a tower that is 150 feet high casts a shadow that is 210 feet long. Find, to the nearest tenth of a degree, the angle of elevation of the sun at that point.   _____

21. A kite is flying 67 meters above the ground, and its string is 90 meters long. Find the angle, to the nearest tenth of a degree, that the kite string makes with the horizontal.   _____

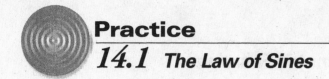

Practice
14.1 *The Law of Sines*

Use the given information to find the indicated side length in
△*ABC*. Round answers to the nearest tenth.

1. Given m∠*A* = 28°, m∠*B* = 95°, and *a* = 12, find *b*. _____

2. Given m∠*B* = 51°, m∠*C* = 70°, and *c* = 30, find *b*. _____

3. Given m∠*A* = 105°, m∠*B* = 64°, and *a* = 18, find *b*. _____

4. Given m∠*B* = 48°, m∠*C* = 62°, and *b* = 25, find *c*. _____

5. Given m∠*C* = 100°, m∠*A* = 82°, and *a* = 5.6, find *c*. _____

6. Given m∠*A* = 75°, m∠*B* = 55°, and *b* = 24.5, find *a*. _____

Solve each triangle. Round answers to the nearest tenth.

7. m∠*A* = 82°, m∠*B* = 60°, *a* = 5

_____

8. m∠*B* = 65°, m∠*C* = 80°, *b* = 20

_____

9. m∠*A* = 100°, m∠*B* = 35°, *b* = 15

_____

10. m∠*A* = 72°, m∠*C* = 64°, *c* = 5.2

_____

11. m∠*A* = 46°, m∠*B* = 52°, *b* = 17

_____

12. m∠*B* = 39°, m∠*C* = 66°, *b* = 54

_____

State the number of triangles determined by the given
information. If 1 or 2 triangles are formed, solve the triangle(s).
Round answers to the nearest tenth, if necessary.

13. m∠*A* = 64°, *b* = 16, *a* = 20 _____

14. m∠*B* = 98°, *a* = 10.5, *b* = 8.8 _____

15. m∠*B* = 28°, *a* = 40, *b* = 26 _____

16. Find, to the nearest tenth of a foot, the length of fence needed to
enclose the triangular piece of land shown in the diagram.

_____

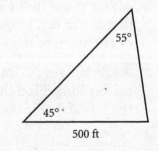

Practice
14.2 *The Law of Cosines*

Classify the type of information given, and then use the law of cosines to find the missing side length of △*ABC*. Round answers to the nearest tenth.

1. m∠A = 46°, b = 24, c = 18

_____

2. a = 25, b = 28, m∠C = 64°

_____

3. m∠A = 32°, b = 8, c = 10

_____

4. a = 10, c = 12, m∠B = 52°

_____

5. m∠C = 110°, a = 16, b = 22

_____

6. b = 8, c = 14, m∠A = 73°

_____

Solve each triangle by using the law of cosines and, where appropriate, the law of sines. Round answers to the nearest tenth.

7. a = 17, b = 15, c = 24 _____

8. a = 50, b = 31, c = 46 _____

9. a = 12, b = 11, c = 9 _____

10. a = 4.5, b = 3.2, c = 6.1 _____

11. a = 10, b = 15, c = 13 _____

Classify the type of information given, and then solve △*ABC*. Round answers to the nearest tenth. If no such triangle exists, write *not possible*.

12. m∠B = 110°, a = 75, c = 85 _____

13. a = 6.2, b = 8, c = 4.2 _____

14. a = 200, b = 100, c = 150 _____

15. m∠A = 50°, b = 3, c = 8 _____

16. m∠C = 95°, a = 8, c = 6 _____

17. Find, to the nearest tenth of a degree, the measures of the angles of an isosceles triangle in which the base is half as long as each side.

_____

Practice
14.3 Fundamental Trigonometric Identities

Use definitions to prove each identity.

1. $\dfrac{\cot \theta}{\cos \theta} = \csc \theta$

2. $(\sin \theta - 1)(\sin \theta + 1) = -\cos^2 \theta$

3. $1 + \cot^2 \theta = \csc^2 \theta$

4. $1 + \tan^2 \theta = \sec^2 \theta$

Write each expression in terms of a single trigonometric function.

5. $\dfrac{1 - \cos^2 \theta}{\sin \theta}$ _____

6. $(\sec \theta - 1)(\sec \theta + 1)$ _____

7. $\cot^2 \theta - \csc^2 \theta$ _____

8. $\dfrac{(\sin \theta)(\cos \theta)}{1 - \sin^2 \theta}$ _____

9. $\csc \theta - (\cos \theta)(\cot \theta)$ _____

10. $(\sin \theta)(\tan \theta) + \cos \theta$ _____

Write each expression in terms of $\sin \theta$.

11. $\dfrac{\csc \theta - \sin \theta}{\cot^2 \theta}$

12. $(\csc \theta - \cot \theta)(1 + \cos \theta)$

13. $\sin \theta + (\cos \theta)(\cot \theta)$

Practice

14.4 Sum and Difference Identities

Find the exact value of each expression.

1. $\sin\left(\frac{3\pi}{4} + \frac{\pi}{3}\right)$

2. $\sin\left(\frac{3\pi}{4} - \frac{\pi}{3}\right)$

3. $\cos\left(\frac{3\pi}{4} + \frac{\pi}{3}\right)$

_____ _____ _____

4. $\cos\left(\frac{3\pi}{4} - \frac{\pi}{3}\right)$

5. $\sin\left(\frac{3\pi}{4} + \frac{2\pi}{3}\right)$

6. $\cos\left(\frac{3\pi}{4} + \frac{2\pi}{3}\right)$

_____ _____ _____

7. $\sin\left(\frac{\pi}{6} + \frac{\pi}{4}\right)$

8. $\sin\left(\frac{\pi}{6} - \frac{\pi}{4}\right)$

9. $\cos\left(\frac{\pi}{6} + \frac{\pi}{4}\right)$

_____ _____ _____

Find the exact value of each expression.

10. $\sin(-285°)$

11. $\cos(-285°)$

12. $\sin 135°$

_____ _____ _____

13. $\cos 210°$

14. $\cos(-75°)$

15. $\sin 345°$

_____ _____ _____

16. $\cos 345°$

17. $\sin 240°$

18. $\sin(-75°)$

_____ _____ _____

Find the rotation matrix for each angle. Round entries to the nearest hundredth, if necessary.

19. 120°

20. 135°

21. 225°

_____ _____ _____

22. 65°

23. −40°

24. 112°

_____ _____ _____

25. A rectangle has vertices at (3, 5), (3, 10), (7, 10), and (7, 5). Find the coordinates of the vertices after a 135° counterclockwise rotation about the origin. Round coordinates to the nearest hundredth.

_____

Practice

14.5 Double-Angle and Half-Angle Identities

Verify that the double-angle identities and the half-angle identities are true for the sine and cosine of each angle.

1. 90° _____

_____

2. 120° _____

_____

3. $\dfrac{\pi}{3}$ _____

_____

4. $\dfrac{2\pi}{3}$ _____

_____

Write each expression in terms of trigonometric functions of θ rather than multiples of θ.

5. $\sin^2\left(\dfrac{\theta}{2}\right)$ 6. $\cos^2\left(\dfrac{\theta}{2}\right)$ 7. $\dfrac{\sin 2\theta}{\tan \theta}$

Simplify.

8. $\dfrac{1 - \cos 2\theta}{1 + \cos 2\theta}$ _____

9. $\dfrac{1 + \sin \theta - \cos 2\theta}{\cos \theta + \sin 2\theta}$ _____

10. $\dfrac{1 + \cos 2\theta}{\sin 2\theta}$ _____

11. $\left(\sin\left(\dfrac{\theta}{2}\right) + \cos\left(\dfrac{\theta}{2}\right)\right)^2$ _____

12. The angle of elevation of a flagpole was measured at distances of 45 feet and 14.4 feet from the flagpole. The second measure of the angle of elevation was twice the first. Find the height of the flagpole. _____

Practice

14.6 Solving Trigonometric Equations

Find all solutions of each equation.

1. $\sin^2 \theta = \frac{1}{4}$

2. $\tan \theta - 1 = 0$

_____   _____

3. $\sec \frac{\theta}{2} = 2$

4. $2 \cos \frac{\theta}{2} - 1 = 0$

_____   _____

Find the exact solutions of each equation for $0° \le \theta < 360°$.

5. $2 \cos^2 \theta - 3 \cos \theta = 2$

6. $2 \sin^2 \theta + 3 \sin \theta + 1 = 0$

_____   _____

7. $4 \cos^2 \theta - 2 = 0$

8. $2 \sin^2 \theta = \cos 2\theta$

_____   _____

Find the exact solutions of each equation for $0 \le x < 2\pi$.

9. $2 \sin \frac{x}{2} - 1 = 0$

10. $\cos x - \sin x = 0$

_____   _____

11. $2 \cos 3x - 1 = 0$

12. $2 \sin^2 x - \sin x - 1 = 0$

_____   _____

Solve each equation to the nearest hundredth of a radian for $0 \le x < 2\pi$.

13. $9 \cos^2 x - 1 = 0$

14. $6 \sin^2 x - 5 \sin x + 1 = 0$

_____   _____

15. The equation $y(t) = 122t \sin \theta - 16t^2$ describes the altitude of a ball t seconds after it was hit at an angle of θ degrees. Determine, to the nearest tenth of a degree, the measure of the angle at which the ball was hit if it had an altitude of 20 feet after 2.8 seconds.

_____